WILD SHORES

About the Author

Richard Nairn is an ecologist and a lifetime sailor. During his career he worked as a nature reserve warden and was the first director of BirdWatch Ireland. His previous book *Wildwoods* recounted his experience of managing a small woodland.

Also by Richard Nairn

Wild Wicklow: Nature in the Garden of Ireland
Ireland's Coastline: Exploring its Nature and Heritage
Bird Habitats in Ireland (joint editor)
Dublin Bay: Nature and History
Wildwoods: The Magic of Ireland's Native Woodlands

WILD SHORES

THE MAGIC OF IRELAND'S COASTLINE

RICHARD NAIRN

Gill Books

Gill Books
Hume Avenue
Park West
Dublin 12
www.gillbooks.ie

Gill Books is an imprint of M.H. Gill and Co.

978 07171 9276 2

Designed by Bartek Janczak
Print origination by O'K Graphic Design, Dublin
Edited by Sheila Armstrong
Proofread by Neil Burkey
Indexed by Cliff Murphy

Printed by CPI Group (UK) Ltd, Croydon, CR0 4YY
This book is typeset in 12.5 on 21pt, Sabon.

A CIP catalogue record for this book is available from the British Library.

5 4 3 2 1

For Wendy

And to the memory of my father,
George E. Nairn (1920–2014)

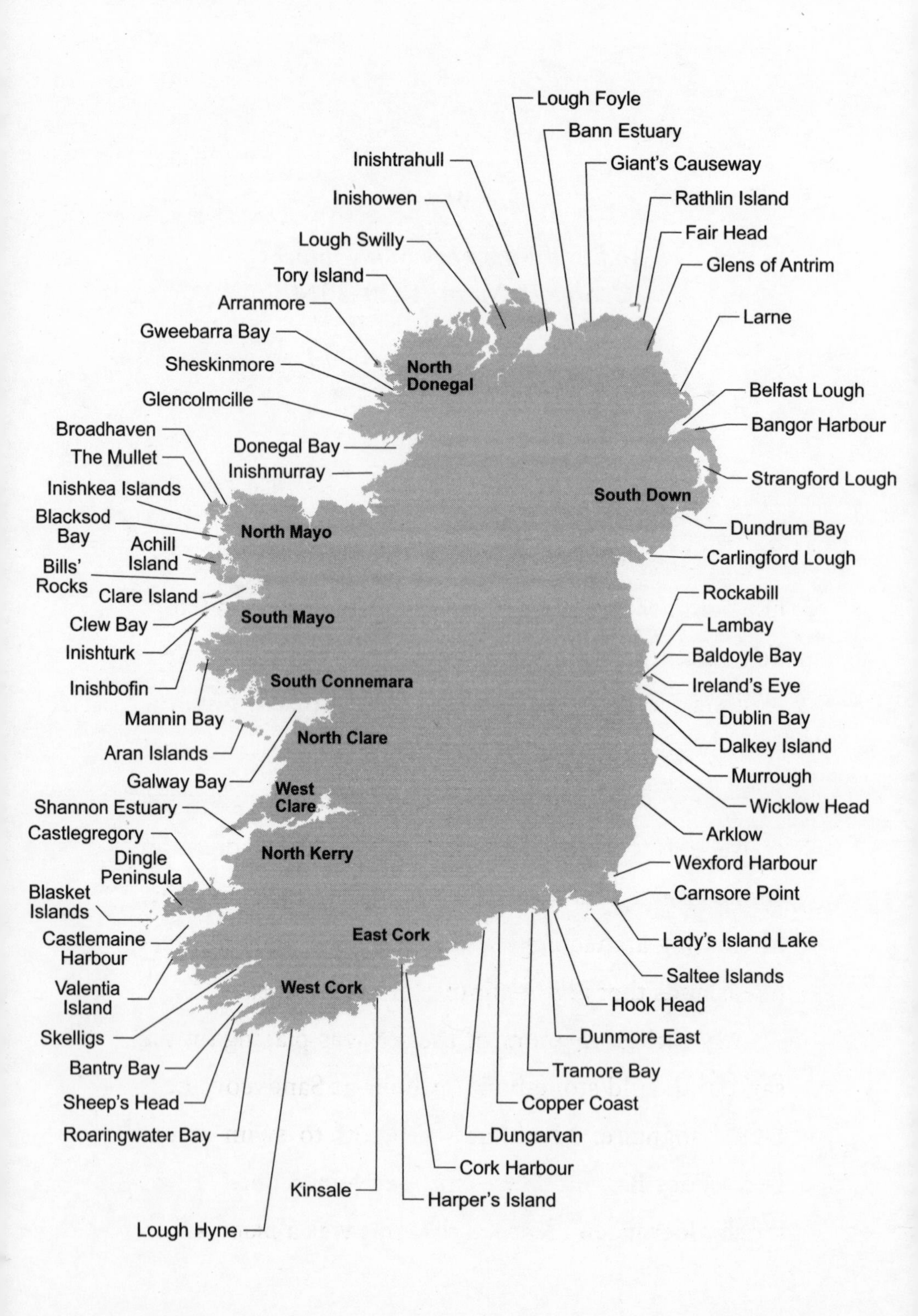
Lough Foyle
Bann Estuary
Giant's Causeway
Rathlin Island
Fair Head
Glens of Antrim
Larne
Inishtrahull
Inishowen
Lough Swilly
Tory Island
Arranmore
Gweebarra Bay
Sheskinmore
Glencolmcille
North Donegal
Belfast Lough
Bangor Harbour
Strangford Lough
South Down
Dundrum Bay
Carlingford Lough
Broadhaven
The Mullet
Inishkea Islands
Blacksod Bay
Achill Island
Bills' Rocks
Clare Island
Clew Bay
Inishturk
Inishbofin
Donegal Bay
Inishmurray
North Mayo
South Mayo
South Connemara
Mannin Bay
Aran Islands
Galway Bay
North Clare
West Clare
Shannon Estuary
Castlegregory
Dingle Peninsula
Blasket Islands
Castlemaine Harbour
Valentia Island
Skelligs
Bantry Bay
Sheep's Head
Roaringwater Bay
Lough Hyne
Kinsale
Harper's Island
Cork Harbour
Dungarvan
Copper Coast
Tramore Bay
Dunmore East
Hook Head
Saltee Islands
Lady's Island Lake
Carnsore Point
Wexford Harbour
Arklow
Wicklow Head
Murrough
Dalkey Island
Dublin Bay
Ireland's Eye
Baldoyle Bay
Lambay
Rockabill
North Kerry
East Cork
West Cork

Preface

To most people, the coast is associated with holidays, hot summer weather, beaches, sand dunes and fun. To some, it is a place for regular sport and exercise – walking the dog, cycling on coastal paths, swimming, sailing, rowing, angling – and many other water sports. For a few, it is their workplace – fishermen, shipping crews, port workers, lighthouse staff. To me, it offers an endless range of interests – a wide diversity of wildlife, fascinating archaeology and history and remote places like islands that give a glimpse of wilderness.

My earliest memory of the sea was playing on the sand in the old stone-built harbour at Sandycove near Dún Laoghaire. My father went off to swim at the Forty Foot Bathing Place 'for Gentlemen only'. I was much older when I learned that this was a place where men were only required to wear bathing costumes after nine o'clock in the morning.

When my earlier book *Ireland's Coastline* was first published in 2005, many people asked me if their favourite place was included. I had to explain that the book was not a geographical guide but instead a general account of the ecology, history and uses of the coast. In response to these requests, I have now attempted to fill that gap. This book follows the Irish coastline from place to place, starting and ending at the north-east corner. I have explored almost all the places mentioned here either by boat or on foot. Occasionally, I have flown over them. Inevitably, my primary interests in ecology and nature conservation are discussed throughout the book, but I regularly divert into the subjects of geology, history and archaeology wherever these are relevant. Nevertheless, there are still some secret places that I have not yet reached, either because they are mostly inaccessible or because it is impossible, in one lifetime, to visit every small bay, headland and island in a tortuous coastline some 7,500 kilometres in length. The selection of places mentioned in the text is a very personal one and it is not possible to include every jewel on this endless chain.

Throughout the book, I have referred frequently to Ireland's greatest naturalist, Robert Lloyd Praeger, who recorded, through his writings, accounts of many of these places a century earlier. Praeger was a prolific traveller and writer and he left behind an enormous legacy of scientific publications, some of them being among the first to describe in detail the flora of Ireland. My text includes many quotations from Praeger's best known and more popular books, *The Way that I Went,*[1] *Beyond Soundings*[2] and *A Populous Solitude.*[3] After the first mention, these books are referenced by (WW), (BS) and (PS) respectively, to avoid repetition in the reference list. In describing Praeger's life I have been greatly assisted by two fine biographies of the naturalist by Timothy Collins[4] and Seán Lysaght.[5] Praeger's writings and adventures 'for seven decades of robust physical health' have been inspirational to me throughout my life and I have tried to follow in his footsteps throughout Ireland on land or on the sea.

Understanding the weather and tides makes me acutely aware of how dependent we are on the sea and gives me a unique perspective on the Irish coast. In the main chapters I follow the coastline in a clockwise direction starting in County Antrim. In the final chapter,

'Turning the Tide', I reflect on some current threats to the coastal environment and how these might be approached in future.

The American writer and biologist Rachel Carson summed up how fundamental the coast is to our well-being. 'Like the sea itself, the shore fascinates us who return to it, the place of our dim ancestral beginnings. In the recurrent rhythms of tides and surf and the varied life of the tide lines there is the obvious attraction of movement and change and beauty. There is also, I am convinced, a deeper fascination born of inner meaning and significance.'

My father taught me to sail in a wooden dinghy around Dublin Bay and I have enjoyed boats all my life. I love to be out on the sea with just the wind, the waves and some good friends for company. I would like my children and grandchildren to enjoy the same privilege.

Acknowledgements

I would like to thank the following for information and guidance: Claire Barnett, Helen Boland, Jason Bolton, John Brannigan, Bob Brown, Amanda Browne, David Cabot, Regina Classen, Andy Crory, Alex Crowley, Marie Duffy, Brendan Dunford, Una Farrington, Pádraic Fogarty, Tom Gittings, Rob Goodbody, Clare Goodwin, Jeremy Greenwood, Julian Greenwood, Bernie Guest, Stephen Heery, Madeline Hutchins, Matthew Jebb, David Jeffrey, Emmett Johnston, Fintan Kelly, Des Lavelle, Cormac Lowth, Patrick Lynch, Seán Lysaght, Emer Magee, Ferdia Marnell, Billy McClelland, Jimmy Murray, Tony Murray, Stephen Newton, Francis O'Beirne, Micheál Ó Briain, Eamonn O'Reilly, Karl Partridge, Graham Roberts, Ralph Sheppard, Dave Suddaby, Michael Viney, Hans Visser,

Philip Watson, Jo Whatmough, Padraig Whooley, Paddy Woodworth, Jim Wilson and Shane Wolsey. I offer my special thanks to Liam Blake, Brian Burke, John Fox and Tatiana Lie Kumagia who allowed me to use their photographs in the book. I am very grateful to Jane Clarke for permission to reproduce her poem 'At Purteen Harbour'.

I am also indebted to those who read and commented on sections of the book. They include Dermot Breen (Connemara), Bob Brown (Ulster coast), David Cabot (West Mayo), Gordon D'Arcy (North Clare), Clare Heardman (West Cork), Jim Hurley (South Wexford), David McGrath (Galway Bay), Declan McGrath (Waterford), Ralph Sheppard (Donegal), Pat Smiddy (South Cork) and Fintan Kelly (fisheries). My special thanks are due to Bob Brown, Matthew Jebb, Cormac Lowth and Karl Partridge, each of whom read the entire manuscript and made many suggestions for improving it. As usual, the editor, Sheila Armstrong, helped to make my text more readable and gave me lots of ideas for additions. Laura King skilfully arranged the pictures. I would like also to acknowledge the help I received from Bernadette Cunningham and Konstantin Ermolin of the Royal Irish Academy who readily gave me access to the Praeger Collection in the Academy's library. The Academy is acknowledged for permission to reproduce one picture from this collection.

I thank my regular sailing companions Cormac Byrne, John Kissane and Brian Malone who have endured, with patience, my rambling commentaries on the places we visited by sea. Robert Lloyd Praeger dedicated his most famous book, *The Way that I Went*, 'to my dear companion in many wanderings'. I too dedicate this book to my wife Wendy who has explored with me many of the most interesting parts of the Irish coast.

CONTENTS

Introduction

The sun was already sinking fast over the coastal town of Wexford when I arrived at Rosslare Point on a fine summer evening fifty years ago. I had been delayed in Dublin and arrived two hours later than planned so my friend David Cabot had already crossed to the island where the terns were nesting. This was before the era of mobile phones so there was nothing for it but to borrow a small wooden rowing boat from a local man and set off across Wexford Harbour on my own. I had no fear as I had been messing about in boats since childhood and I have a strong rowing stroke. But I had underestimated the tidal currents. Twice each day a vast body of water fills and empties through the narrow mouth of Wexford Harbour and, though shallow, the currents can be faster than a river.

As I reached the centre of the channel in the fading light, I realised that I was being dragged by the tide further and further out into the Irish Sea and away from the sandy island. It crossed my mind that I had no radio or way of contacting help. My foolishness became crystal clear when I remembered that I did not even have a lifejacket. I had only a wooden boat and two heavy oars. For the first time in my life, I tasted that strange mixture of fear and respect for the power of nature. My survival was now a toss-up between the sea and my own resources. Then my years of training kicked in. Summoning all my strength and stamina, I pulled hard on the oars and turned across the tide. Fortunately, after more than an hour of exhausting rowing, the tide turned, and I was able to pull the boat up on a shelly beach and sit down to recover. Thoughts of Robinson Crusoe passed through my mind as the small island was uninhabited. I lay back on the sand and counted my blessings.

The turn of the tide is one of those immutable things in life. Like night following day, summer fading to autumn, middle age after youth, I know it will always come. I cannot stop its steady progress. It never lets me down. Ebb and flow, fall and rise, twice a day, unfailing, dependable, predictable. In a crazy, uncontrollable world, I know that, as long as the moon is in the sky, the tide will never fail. It carries my boat along or stops it in its tracks. It pushes up the beach, slowly erasing my footprints. It leaves behind rich offerings. A spiny spider crab that once crept about the deep seabed, a heap of oyster shells dredged from the sandbanks offshore or the bleached skull of a seabird that did not survive the winter.

When the tide turns against the wind direction, it pushes the sea surface up into small crests that change the motion of a sailing boat, like driving from the road onto a gravel track. Just as the wind moves the air about, tides pull and push seawater around the coast, in and out of rock pools, up and down a beach. But unlike the wind, which is invisible, unpredictable, sometimes gentle, often angry, I can see the tide as it passes, watch its steady progress, make allowances for its effects.

The turn of the tide is a wondrous thing. The constant movement of the ocean pauses for a moment, takes a deep breath, reaches its limits and starts to return the way it came. Curtains of seaweed are swept in the opposite direction, sand grains roll down the beach, and crabs scuttle into deeper water. I feel a different pull on the boat as it implores me to follow rather than resist it. The tide is my constant friend, not a threatening adversary. But I have reached this accommodation after fifty years of experience of the sea. Sitting on that beach, it seemed like the tide had granted me a narrow escape and taught me a lesson for the future.

Tern Island in Wexford Harbour was then little more than a sandy ridge with a thin covering of dune grasses but at the time it held one of the largest colonies of terns in Ireland. Five different species of these small, delicate seabirds arrived to breed here each summer from their African winter quarters. The nests were closely spaced and little more than shallow scrapes in the sand, sometimes lined with a few shells or blades of grass. My friend David, who had travelled to the island ahead of me, had invited me to help him with a long-term research project on the rare roseate tern that nested on Tern Island in greater numbers than any other place in Europe. I was thrilled to be involved with a project where I got to see the birds at close quarters. This research had a practical application for conservation, as fitting a small sample of the terns with tiny numbered leg rings in Ireland demonstrated that these migratory birds were being hunted in Africa and the population was in steep decline. Alas, a few years later, the island was completely destroyed by a series of winter storms and the terns were forced to find other breeding sites.

As a student at Trinity College Dublin, in the early 1970s, I was excited by the idea that there were wild places and species out there somewhere waiting to be discovered. I read avidly the works of the old naturalists like Robert Lloyd Praeger, hoping to re-find some remnants of the pre-industrial Ireland that they described. Praeger began *The Way that I Went*, his book about a life of wanderings in Ireland, with the words: 'It is indeed a kind of thank-offering, however crude, for seven decades of robust physical health in which to walk and climb and swim and sail throughout or around the island in which I was born, to the benefit alike of body and soul.'

Praeger was seventy years old when he wrote his famous book. Over the course of my own seven decades, I have often thought of Praeger as we crossed the same paths, although he died the year after I was born. He began life near Belfast and I was born near Dublin. In his twenties, he moved to Dublin for his first permanent job while, at the same age, I moved to Northern Ireland to work. My second job was in the Royal Irish Academy in Dublin where Praeger had been President fifty years earlier. In the Academy's library, I was fortunate to be able to view the collection of Praeger's own papers including a signed copy of his ground-breaking research, the Clare Island Survey. Although I will never match his achievements, he has been an inspiration throughout my life which has been enriched by natural history, reading and writing. But, in the near century that separated us, Ireland changed almost beyond recognition.

In my search for 'wild' Ireland, I was to be disappointed again and again as I travelled around most of the island. The native forest cover of Ireland had been removed thousands of years earlier. Mechanised farming and forestry had destroyed much of the wildness that Ireland contained in the 19th and early 20th centuries. The landscape was now highly modified from the seashore to the summits of the highest mountains. It seemed virtually impossible to find any places that were untouched by people and their use of the land.

But some small wild areas remained on the salt-sprayed coasts and islands and in the untamed region between the tides where the land and sea overlap. I found that these were the places to look for the best of Ireland's wildlife. Even walking the wide beaches of Dublin Bay, within sight of the city centre, I experienced a closeness to nature and the natural energy of tides, winds and waves. Slowly, I came to accept that the long history of human activity on the coast was as much a part of our heritage as the wilder areas that seemed to have survived untouched. So for this book, I have given equal attention to the landscape, history and ecology of the coast in the hope of presenting a balanced and realistic appraisal of the most interesting places.

Seamus Heaney once described the coastline as a place where 'things overflow the brim of the usual'. But what exactly is the coast? Many experts have tied themselves into knots trying to define what constitutes the coastal zone, although everybody knows that this is where we find beaches, waves, seaweed, shellfish and seabirds. The core of the coastal zone is the area between high and low water mark that is covered and uncovered by seawater twice each day. Inland of this are coastal lands such as sand dunes and cliffs that are strongly influenced by marine processes, while to seaward are shallow marine areas that receive sediments from rivers and recycling of beach sands.

Estimates of the length of the Irish coast vary greatly depending on whether the perimeters of every minor rock, peninsula and inlet are included. The best estimate is one based on Geographic Information Systems that arrived at a figure of 7,524 kilometres.[6] Nearly half of the total length of coastline is found on the highly indented west coast between Cork and Donegal, now promoted as the 'Wild Atlantic Way'. The counties with the longest coastlines in Ireland are Cork, Kerry, Clare, Galway, Mayo and Donegal. The largest proportion of the Irish coastline is made up of rock, sand and mud but there are substantial sections with artificial sea walls and harbours. We are essentially a maritime people, as nowhere in this island is more than 100 kilometres from the coast. In 2016 there were 1.9 million people living within five kilometres of the Irish coast, representing 40 per cent of the total population (Central Statistics Office, 2017).

I have always lived close to the coast. As a young man, my father was a naval officer and he remained involved in sailing all his life. He taught me the skills of rowing, sailing, swimming and fishing and he also gave me a love of the salt sea air and the freedom that comes with it. Since I was a child I have been sailing in all sorts of boats, from single-handers to large ocean-going yachts. I get a sense of exhilaration and anticipation when casting off from any harbour. Out on the waves I leave behind the cares and complications of everyday life, focusing just on the weather, the tide, the boat and my companions. I love to walk along the tideline on a western beach, searching the flotsam and jetsam of the oceans for unusual shells, seaweeds and the bleached white bones of dolphins. My grandfather's house was in Sandymount, just one street away from the shoreline of south Dublin Bay. He told me that, when he dug a hole in the back garden, his spade went straight into beach sand full of seashells. My great-grandfather was an artist who lived on the northern shores of Dublin Bay, sketching and painting the sailing ships and fishing craft that visited the port in large numbers in the nineteenth century. I have saltwater in my veins and I feel uncomfortable living away from the sea.

In my college days, when I was staring out the window of the library, I dreamed of living and working in a wild place close to nature and learning first-hand how the natural world works. My first proper job, when I emerged into the sunlight after four years studying natural sciences at university, was on a nature reserve in Northern Ireland where I was appointed as one of the wardens. The place is called Murlough – a wonderful area of sand dunes with a stunning backdrop of the Mourne Mountains sweeping down to the sea. Here was the ideal opportunity and I couldn't believe my luck. I caught the train from Dublin to Belfast and arrived at Murlough with a bag on my back and excitement in my head. This was the early 1970s and Northern Ireland was in a deep political crisis with armed conflict and increasing militarisation. Within days of my arrival, Northern Ireland was plunged into darkness as the Ulster Workers' Council Strike closed down the main electricity-generating station. Before long, the period known as 'The Troubles' became just part of daily life and I got on with the work to which I had been appointed.

The accommodation was fairly spartan. I spent the first summer in an abandoned cottage with no electricity or running water, using rainwater that drained from the roof into a barrel. It was right on the edge of the estuary and I could hear the curlews and oystercatchers calling from the shoreline through the night. Early one morning, when I looked out through a cracked windowpane, I saw an otter loping along the edge of the tide stopping occasionally to leave its mark. In August that year, when an unusually high tide started to lap at the door of the cottage, I decided it was time to move on. I packed my rucksack and cycled over the sand dunes to the big house near the beach. This was owned by the Marquis of Downshire, but his lordship was not in residence. The land had recently been transferred to the National Trust and the big house now lay vacant. In an old stableyard nearby lived an elderly couple who had worked for the estate in its heyday. They kindly offered me an upstairs room in the stables as a temporary home.

With winter coming on I made myself as comfortable as possible. I found an old rusting iron bed in a shed, borrowed a mattress and a kettle and started to cut up some firewood to keep the place warm. Although I didn't know it then, this stable was to be my home for the next five years. Despite the lack of home comforts, it was an inspiring place for a young naturalist to begin his first real job: living in the centre of a nature reserve just a hundred metres from the beach. From the draughty windows, I could hear the waves on the shore and, when the south wind blew, the calls of seals hauled out on the bar were plainly audible. This plaintive sound could be easily mistaken for mermaids singing to attract humans to their watery home. There were many other unfamiliar sounds as well. Herons roosted in the pine trees around the house and their night-time squawks would waken me from sleep. I heard the evocative calls of curlews and wild geese flying overhead. As the winter progressed the trees filled with flocks of thrushes, many of them migrants from Scandinavia or Iceland, and the calls of these birds going to roost at night will stay with me for ever.

My appetite for working and living in remote wild parts of the coast had been whetted by two main influences. I had read avidly the works of Frank Fraser Darling, who farmed a ruined croft on the deserted Isle of Tanera in the Scottish Hebrides. He was a superb naturalist and used his time in the wild to make some ground-breaking zoological studies of seabirds, seals and deer. I treasured his classic book *Island Farm* which was published in the 1940s on the cheap, low-quality paper that marked the war years in Britain.[7] His many later books revealed his own passionate character as well as the science that drove him to these wild places. Fraser Darling went on to become a charismatic if always controversial figure in the world of international conservation. His adventures captured my imagination and I was fortunate, like him, to live and work close to nature.

While I was a warden at Murlough I was lucky to take part in several expeditions to study seabirds. On the Saltee Islands off Wexford we stayed in an old farmhouse, collecting water in buckets from the well. Health and safety were minor considerations here as the team of enthusiastic young ornithologists scrambled among the cliff ledges and boulders counting fulmars, gannets and puffins. We used a strange contraption like a giant butterfly net to swipe through the air and catch guillemots and razorbills as they were startled from their nesting ledges. These birds were caught unharmed, fitted with metal leg rings and released in the hope that they would be found again on some distant coast or, better still, by another ornithologist studying the same species. This was a formative time and it gave me a clear direction for my life.

Shortly after this, my friend David came to visit me at Murlough where I was now well settled into the stable quarters. With our telescopes we spent hours on the beach searching through the seabird flocks that gathered on the beach, looking for coloured rings that might provide a hint of their natal colony. One morning, without warning, gunfire rang out as bullets ripped through the vegetation around us. We didn't wait to ask questions but dived back into cover leaving behind our telescopes which, in retrospect, could have been mistaken for automatic weapons. The sand dunes across the channel were used by the British Army as a training ground and the shots probably came from a young trigger-happy recruit who thought we were part of the military exercise.

Luckily, I survived these early experiences and began to enjoy life as a warden-naturalist. I learned so much in those years about natural history in coastal habitats but also about the business of managing land for conservation. I met some eminent people involved in coastal research and conservation such as Bill Carter and Palmer Newbould. Thirty years later, I wrote a book called *Ireland's Coastline*, to put down on paper what I had learned and to make a plea for the nature and heritage of our maritime fringe.[8] Since then, much has changed. The climate is becoming warmer each decade and this has serious implications for seawater and coastal habitats. Overfishing is having increasing impacts on the marine ecosystem. Plastic pollution of the oceans has become a major issue.

International travel restrictions in the wake of the Covid-19 pandemic have forced Irish people to explore and appreciate their own beautiful country. This book is a guide to the best parts of the Irish coastline but it is a distinctly personal selection. With over 7,500 kilometres of shoreline, including all the islands and bays, to choose from, I could only pick those places that I know well and give just a sample of the richness and diversity that the Irish coastline holds. The foreshore – the land exposed when the tide drops to its lowest each day – is owned by the state in the Republic, by the Crown Estate and local authorities in Northern Ireland and by the National Trust in the case of Strangford Lough. From the vast tidal mudflats of the Shannon Estuary to the highest rocky cliffs in Europe, it is free to roam and to explore.

As Robert Lloyd Praeger wrote in his classic autobiography *The Way that I Went*, 'I have done what I said I would, and roamed at random.' I often take his book with me on my own travels to different spots around the Irish coast and consult it to see what he found there a century earlier. His memoirs are filled with vivid accounts of the landscape, its flora and fauna, historical and archaeological features, Irish placenames and, unlike most nature writing, frequent quotes from his favourite poetry. In one of his later books, *A Populous Solitude,* Praeger included memories of many people that he met on his travels:

> I strolled downhill towards where the sound of steel on steel told of the presence of human industry and came upon a man engaged in driving a line of wedges into a granite block. Near-by lay the result of previous labour – rough six-foot squarish pillars, to be used evidently as gate-posts. On one of them a woman rested with a basket and 'billy-can', waiting for the completion of the work to offer him tea and food.

His penultimate book, *Some Irish Naturalists*, includes sixty short pen portraits of both amateur and professional natural scientists, including members of his own family, that he knew during his long life. He wrote:

> I think my earliest contact with a man of science was when my grandfather, Robert Patterson, took me to Cultra on the shores of Belfast Lough to show me the Adder's-tongue. By inclination he was a zoologist, but in those days, naturalists were not specialists, and with a wide knowledge of the animal kingdom he was also well acquainted with local plants.[9]

The chapters of my book that follow this introduction have been written following Praeger's example on a rambling journey of memories around the coast of Ireland in a clockwise direction. Since my father undertook a complete circumnavigation of Ireland by yacht, I have always wanted to make a similar journey by sea. So far, however, I have only managed to sail the east and south coasts. I am lucky to have visited by land many parts of the west coast mainland and islands while I explored much of the north coast on foot. My yacht *Ventosa* is a 32-foot sloop with four berths. In it there is a collection of books including those of Praeger and I always find something of interest here during my travels. I have visited all the places mentioned in the text but at different times in my life and using different modes of transport. I start this journey on a yachting passage in the north-east corner of Ireland just a few short miles from Belfast Lough where Praeger began his life and his ramblings.

East Coast

Robert Lloyd Praeger was born in Holywood, County Down in 1865. He was, by all accounts, a bright child and the earliest photograph shows him at six years old reading a book while seated at an intricately carved desk. In his own words, 'at school I was idle and inattentive; I hated schoolbooks and school-masters and made some approach to being a dunce'.(PS) He came from a stoutly Protestant tribe and, though not religious in the conventional sense, the preface to his most famous book *The Way that I Went* contains the line 'Thank God for Life'. He was described as gruff in manner but a kind and gentle soul throughout his life. His father was a Dutch Presbyterian who had emigrated to Belfast in 1860, there married into the Patterson family and became involved in the linen business for which Ulster was then well-known.[1] It was to his mother's family that Robert Lloyd Praeger ascribes his early interests in natural history, through his grandfather and especially his uncle Robert Lloyd Patterson. With his sister and three brothers he wandered around the nearby Holywood Hills, exploring further and further afield as he grew.

It must have been a privileged childhood – loving parents and grandparents, a private school, all the books that he could read – and he benefitted from this in his education. After graduating from Queen's University Belfast, he qualified as an engineer and had various short-term contracts working on construction projects in the northern counties. But all the while he was honing his skills as a naturalist and a writer. As a young man, he spent his leisure time with his brothers and friends walking and exploring the hills of Ulster, of which the Mourne Mountains were a particular favourite. Among his many adventures one stands out for its drama. Having walked all day alone from the hills of Slieve Gullion to the west he sat among the heather of the Mournes to rest and dream of his earlier life.

> A distant hallo brought me to my senses. Nothing moved in the valley save a couple of distant

> sheep; but looking down, I saw my brother standing in the sunlight in the Gap. He had come by the mountain route from Newcastle – over the 1,900 foot col between Slieve Donard and Slieve Commedagh and on by the sheep-path along the back of Slievenaglogh. Years later, after my brother's death, I recalled that scene – the dark rocks still in shadow, and the erect figure in front of them bathed in bright light.(PS)

Despite being fit and healthy Praeger does not appear to have had any interest in sport or other competitive games. His friend Anthony Farrington wrote, 'Nonetheless he was athletic in the sense that he was a remarkable walker. Not many years before his death he came across an old diary which he showed to the writer. In it was a record of a walk he had taken up one of the Glens of Antrim to meet the Belfast Field Club on the plateau. He had checked his time on the milestones and the average time was less than 10 minutes per mile – one mile was done in 9 minutes.'[2]

Glens of Antrim

The Antrim Glens lie just a short drive to the north of

Belfast and they were a favourite destination of Praeger and his brothers in their youth. These are deep valleys through the relatively level plateau of County Antrim which was formed over fifty million years ago when molten lava forced its way up from the hot centre of the Earth and cooled between the layers of rock above it. Due to the rapid cooling of the lava, it shrank and cracked, with some of it forming the hexagonal columns that are so well known today at the Giant's Causeway. In subsequent eons, the younger rocks on top were eroded off by millions of years of ice and rain, exposing the harder volcanic basalt below.

My journey starts at the little seaside town of Cushendun just eight kilometres south of Torr Head which marks the north-eastern corner of Ireland and from which the south-west coast of Scotland is so close that the individual houses are visible on a clear day. My yacht was moored overnight in the bay just off the beach, where it was well sheltered from the western winds. It was soon after dawn when a big red sun broke above the horizon and lit up the cliffs with a touch of rouge. Praeger described Cushendun 'with its sheltered bay and curious series of caves cut in "pudding-stones" of the Old Red Sandstone'.(WW)

With the wind rising from the north-west it was time to go, so I lifted the anchor and headed out into the Irish Sea. Sailing south, with the outline of the Scottish coast on my left side, I was passing through hundreds if not thousands of Manx shearwaters all flying north as they headed out from their breeding colonies to feed for the day. Their distinctive flap-and-glide flight behaviour marks these black-and-white seabirds out from the gulls and gannets. There is a colony of shearwaters on the Calf of Man and a much larger one on the Copeland Islands off the mouth of Belfast Lough.

From the Irish Sea I had a wonderful view of the Glens of Antrim, deep valleys cut into the edge of the basalt plateau. Looking upwards I could see the distinctive black basalt overlying a thick layer of white chalk in the cliffs looking just like the dark chocolate top on a cheesecake. Praeger walked all of these glens a century ago. He observed that 'the basalt weathers into a heavy rich soil and the narrow, sheltered valley bottoms have much good agricultural land, though on the slopes it soon gives way to gorse and then to heather'.(WW) Praeger loved to meet and talk to local people on his travels. Of 'the Glynns' he writes, 'the glensmen too are worth meeting – a fine hardy, hearty race, closely akin in descent and language to their Scottish neighbours, shrewd, friendly and hospitable'. (WW) I met a few of these hardy folk when my yacht was tied up in the harbour at Glenarm at the seaward end of one of these glens. Among them was Billy McClelland, who runs the small marina here. He told me that in an earlier life he was a deep-sea fisherman. His half-decker fishing boat was tied up just inside the harbour wall. In his youth he would often join the trawlers that fished the Clyde, just across the North Channel in south-west Scotland. Here they would have shelter from Atlantic storms whatever the wind direction. There was much intermarrying between the people of Antrim and south-west Scotland and they still share similar words in their dialects. Today, Billy is content to welcome visiting yachtsmen to his home county of Antrim.

During the 18th and early 19th centuries, Glenarm grew from a small estate village into an important port, exporting great volumes of chalk, iron ore, timber and fish to markets in Scotland and beyond. By 1908 steamships were loading 700 tonnes of iron ore here each week. The sailing ships used were mainly schooners crewed by three men and a boy. From the 1920s these were replaced by small steamships, with up to five of these boats tied up in the harbour at one time loading stone and several more at anchor outside awaiting their turn.

Leaving Glenarm, my friend John and I set sail on a south-east heading with the target of some tiny islands eight kilometres offshore. The Maidens, or Hulin Rocks, are a small group of skerries with a lighthouse on each of the two largest islets. For most of the 19th century, both lighthouses were operational. However, in 1903 the West Maiden was abandoned and the current lighthouse on the East Maiden has been the sole light ever since. Bill Long wrote about the lighthouse keepers and their families who lived here over a century and a half. He imagined all the men, women and children living in the cramped quarters on a tiny patch of land that was often covered by waves. 'There was no room there for the children to run or play or even hide.'[3] Landing here is difficult even in calm weather, and when sea conditions were rough the keepers were often isolated for weeks on end. The Maidens Lighthouse was automated in 1977 and the keepers left the islands for the last time.

There are old records of birds found here by Charles Patten, who published a series of short notes on migrants recorded at the lighthouse during the years of the First World War. These included waders such as redshank and lapwing, songbirds such as mistle thrush, redwing and wheatear and even the tiny wren and quail. Patten was born in Dublin and studied at Trinity College but was appointed Professor of Anatomy at Sheffield University at the early age of thirty-one. Despite this he carried on a series of ornithological studies in Ireland right up to his death in 1948.[4] From the previous century it was known that migrant birds were attracted by the powerful beams of lighthouses at night and often died when striking the structures. He wrote in 1914:

> On Tuesday night, March 31st, Redwings and Fieldfares appeared in large numbers very close round the lantern. Very few actually collided with the glass and when they did so, generally glanced obliquely or, after striking, backed away for a foot or so and then bumped against the glass repeating this performance several times before leaving altogether. At 12.20 a.m. I picked up a

> Redwing and at 1 a.m. a Fieldfare. Both these birds struck the glass and fell wounded on the balcony.[5]

Usually, it was the lightkeepers who collected the dead specimens and posted them to Patten for identification. Given its location between Ireland and Scotland, there would have been a stream of migrant birds passing the lighthouse in spring and autumn, and this almost certainly continues to the present.

The waters around the rocks are very deep, with 180 metres the maximum recorded on my depth sounder. I saw a few storm petrels flitting about at the calm water surface around the islands. These dainty seabirds, about the size of a thrush, are true oceanic birds that only come to land for a short time each summer to breed. Their Irish populations are poorly known, because they nest in crevices in rocks and stone walls and are very difficult to census. Since being abandoned, the West Maiden Rock has become home to a small seabird colony of shags, great black-backed gulls and black guillemots (tysties). Considering their age and the rigours of a century of storms, the old buildings here are in remarkably good condition, and some birds have been nesting inside since the old lighthouse was abandoned. A century's worth of rotten fish and seabird guano (droppings) creates quite a unique, pungent aroma.

Larne

The shortest ferry crossing of the Irish Sea to Cairnryan in Scotland starts in Larne Lough, County Antrim, a sheltered inlet which is almost completely land-locked except for the narrow entrance at the northern end. About halfway between the busy Port of Larne to the north and the shallow estuary to the south are two tiny islands. The smaller of these, Swan Island, is a natural feature comprised of boulder clay, stone and shell material, while the larger island is an artificial construction. As far back as the 1970s, when I was working as a nature reserve warden in Northern Ireland, the project to construct the second island was started by the Blue Circle company, which operated a nearby quarry at Magheramorne. Working with the advice of the late Dinah Browne, then Director of the Royal Society for the Protection of Birds (RSPB) in Northern Ireland, the company used locally available materials, basalt stone, seabed dredgings and kiln dust to build up an island about half the size of a football pitch.

Following construction, the first breeding seabirds on Blue Circle Island were a handful of black-headed gulls. In 1994, the RSPB took on a lease of the island and shortly afterwards, Ireland's first recorded breeding Mediterranean gulls nested there. By 2016 there was a staggering total of nearly 4,300 pairs of seabirds breeding on the artificial island, including the largest individual colony of black-headed gulls and the second-largest colony of sandwich terns in Ireland. The guano produced by the seabirds over the years has now resulted in a lush vegetation, mainly couch grass, which has caused a reduction in numbers of common terns due to decreasing areas of open ground. Ongoing management by RSPB is aimed at recreating the open areas to benefit the terns. Almost fifty years after its inception, the restoration of Blue Circle Island was completed in 2018 to give more space for breeding seabirds. Shane Wolsey, formerly Northern Ireland Officer of the British Trust for Ornithology, says, 'this is a timely conservation project, needed to maintain the structure of Blue Circle Island'. This small site is already a key link in the network of managed sites for threatened terns in Ireland.

Leaving Larne Lough, we sailed close by the cliff-bound Isle of Muck, which lies just off the Antrim coast. The island probably gets its curious name from the Irish word *muc*, meaning pig, which may be because the island was linked with the high number of sightings in this area of the harbour porpoise, known in Irish as *muc mhara* (sea pig). This cetacean sometimes gives a snort just like a pig on surfacing. This impressive island nature reserve, in the care of Ulster Wildlife, contains the third-largest colony of cliff-nesting seabirds in Northern Ireland. Kittiwake, guillemot, fulmar and razorbill all breed here, and peregrines commonly hunt over the island. Most of the Irish gull species are present and the number of common gull nests has been growing steadily in recent years. The populations of breeding seabirds on the island are part of a larger colony that also uses mainland cliffs at the Gobbins.

The Isle of Muck has recently been colonised by eiders. These large ducks are mainly found in arctic waters, but the population also spreads south around the Irish coasts. In the breeding season they make a long 'ooooing' call that echoes across the water. They build deep nests using soft down plucked from their own body (this was traditionally collected for filling eiderdown quilts). But these ground-nesting ducks are very vulnerable to mammal predators and the Isle of Muck was infested with brown rats which had probably made the crossing via the spit that joins the island to the mainland at low tide. The productivity of fulmars on the island was among the lowest in Northern Island, as many eggs and chicks were taken by the rats. A rat-control programme was set up by Ulster Wildlife in late 2017 using a grid of bait traps with rodenticide and sardine oil. This has to be repeated each year until monitoring shows no further rats present. However, they always make their way back gradually from the mainland. Black guillemots, which nest in rock cavities, seem to have benefitted from the absence of rats here, as they have recently increased to over sixty birds. Extensive bracken control has also been completed and grazing by hardy sheep introduced to maintain the open habitats needed by breeding seabirds.

Sailing along below the basalt cliffs, I could see Black Head Lighthouse and the remarkable coastal path known as the Gobbins. Hanging from basalt cliffs directly over the Irish Sea, the Gobbins cliff path was a popular curiosity during the Edwardian period. Designed in 1902, the path was part of the vision of

Berkeley Deane Wise, Chief Engineer of the Belfast and Northern Counties Railway Company, to use the recently expanded railway network to attract visitors to this spectacular part of Ireland. With the arrival of the steam train in the 19th century it became possible for the first time for ordinary people to move about the country. Not only did the railways serve cities like Belfast which were then industrial powerhouses, but they opened up remote beauty spots to a new kind of industry – tourism. In the early 20th century, the old path at the Gobbins was a fee-paying visitor attraction. Each day in the summer season ticket collectors, like retired railwayman Sam Cuthbert, sat in front of an arch cut through the rock which was called Wise's Eye after Berkeley Deane Wise. If you had come by train, he would check your railway ticket. If you came under your own steam, you paid a fee. A gate across Wise's Eye kept people off the old path out of hours.

Praeger was a regular visitor to this part of the Antrim coast, which was a short train ride from his home outside Belfast. Aged 20, he wrote in 1885:

> The first time I paid my respects to Black Head was on a calm evening in August, when my

younger brother and I, having an hour to spare before train time, thought we would go round the base of the cliffs, as it was low water. We got on very well for a while, scrambling over rocks and boulders, and occasionally wading round a projecting point; but presently we came to a deep gully, about 15 feet wide, which ran up into a large cave in the cliff, at the upper end of which the waves were breaking with a hollow roar. However, we got over this obstacle by placing stones in the pockets of our clothes and throwing them across, and then swimming after them. Going on we soon came upon another gully similar to the first, when we repeated the performance. Again proceeding, we came to a third, and before we got clear of the cliffs we had had four or five swims, which, as each involved at least a partial dressing and undressing, detained us considerably longer than we had time to spare, and we missed our train; luckily it was not the last one.[6]

The Gobbins closed in 1954 due to the Great Depression of the 1930s and lack of materials after the Second World War. However, after extensive renovations, the Gobbins was opened again recently as a tourist attraction along the Causeway Coastal Route. It comprises a series of steel bridges along the cliffs, through rock arches, across wave-cut gullies and over cave entrances. While tied up in Larne I walked the five-kilometre route and was amazed by the power of the sea below. I was mesmerised standing on steel bridges above the churning waves, running my hand along the cool stone of the cliff face and enjoying close-up views of many sea birds.

Belfast Lough

A century and a half after Praeger grew up on the shores of Belfast Lough, I made a return visit to the area around the city's busy port, although I had known it well during my time in Northern Ireland in the 1970s. In the 19th and early 20th centuries this was a place of bustling industry. The famous Harland and Wolff shipyards were the site for the building of some of the largest vessels of their time, including the ill-fated *Titanic*. This time I wanted to see how wildlife was surviving within the port, with all its noise, lights and industrial activity. Belfast Lough Reserve is made up of four sites – Belfast's 'Window on Wildlife', Harbour Meadows, Holywood Banks and Whitehouse Lagoon – all of which are special places for nature. These sites are all managed by the RSPB as refuges for birds and other wildlife in the urban environment.

Belfast Harbour Estate can trace its origins back to 1613, during the reign of James I, when a quay was constructed where the River Farset and River Lagan met and the development of Belfast as a port city began. Fast forward to the 1970s, when silt was dredged from shipping channels to allow a more suitable depth in which modern ships could operate. This was pumped ashore, where it settled and hardened with large pools of water – the perfect spot to stop and feed for migrant birds flying overhead. The conservation potential of this site was quickly recognised, and it was designated as a nature reserve in 1998. More than a hundred bird species have now been recorded at the site, from arctic terns to bar-tailed godwits, and the occasional rare visitor is seen from time to time too.

The lagoon at Belfast Harbour is managed for wintering waders and wildfowl, including redshank, oystercatcher and wigeon. The work of the RSPB wardens includes maintaining water levels, mowing grassland and cutting back invasive plants. Water levels in the lagoon here are carefully managed to create the right habitat for all the species that make their homes here. Artificial floating islands give common and arctic terns a safe place to breed. By managing the reedbed the RSPB encourages lots of insects, a valuable food source for species like the sedge warbler, which migrates from Africa in the summer. The resident Konik ponies graze the reserve, creating ideal conditions for wintering wildfowl and ground-nesting birds like lapwing.

I went into the state-of-the art visitor centre which overlooks the reserve and doubles as a birdwatching hide. Major renovations of this centre now give better views of the wildlife on the reserve. I was able to use binoculars and telescopes provided by friendly volunteers who were on hand to help. Other features built to accommodate breeding birds include a sand martin bank and a swift tower. Meanwhile two new hides, which have been constructed from shipping containers to tie in with their surroundings, offer different perspectives on the reserve.

Holywood Banks are among the last remaining mudflats of the many which once surrounded Belfast Lough but have since been filled in. The surviving mudflats are an important habitat for migrating birds like curlew and oystercatcher, which stop here to feed on the long journeys to and from their northern breeding grounds. The RSPB wardens also manage the nearby mudflats at Whitehouse Lagoon for wintering wildfowl and waders, and are working to safeguard both from illegal bait digging and fly-tipping. When the tide goes out Whitehouse Lagoon becomes crowded with wading birds like black-tailed godwits, as they probe the mud in search of food. When the tide comes in many of the birds move across the lough to the Belfast Harbour lagoons.

Bangor Harbour

At the south-eastern end of Belfast Lough lies the seaside town and large harbour of Bangor, County Down. This has been a refuge from storms for me on several occasions, and I am always happy to see it from the sea. A few years ago, I was sailing on this coast with my friend Brian when a storm blew up and the Irish Sea became very choppy. As it turned out, our yacht was tied up in Bangor for three days waiting for the storm to abate so I got to know the harbour well. Walking up the pontoons to the harbour wall I was greeted by the high-pitched calling of black guillemots which seemed to be everywhere. Known as 'tysties' here, as they are in Scotland, this name probably derives from an Old Norse word, *beisti,* brought to Ireland by the Vikings. These small seabirds have a very distinctive black-and-white plumage and bright-red feet set well back on the body. They are relatives of the puffin and, like them, nest in dark cavities, where they lay a single egg. In many harbours around the Irish coast the birds find suitable holes in old quay walls and jetties that mimic the rocky cavities they use in natural cliff sites. Julian Greenwood, then a lecturer in biology at Stranmillis College in Belfast, began to study the birds at Bangor in 1985. Julian happened to live in the town of Bangor, so it was easy for him to walk regularly around the harbour and check how 'his' birds were doing.

Black guillemots began nesting in the old North Pier at Bangor in 1911, using small crevices in the decaying wood and concrete structures. By the mid-20th century, there were probably no more than about six pairs. In the late 1970s the harbour authorities decided to rebuild the old North Pier and add new concrete quays for the fishing boats. This destroyed a number of nest holes used by the birds, but Julian persuaded them to install new wooden nest boxes under the piers and to experiment with adding short lengths of plastic pipe strapped to the vertical sides of the walls.[7] The guillemots took to these new nest sites with enthusiasm, and from then on the population in the harbour grew till it reached a total of thirty-eight pairs in 2013. The birds are very faithful to their individual nest boxes, returning each year to breed.[8] While the population here is in a very healthy state, the breeding success has dipped slightly, and there is some evidence that this may be linked to long-term increases in sea temperature, which may affect their main prey species – the butterfish.

The birds had become such a feature of the harbour, seemingly undisturbed by the frequent movements of boats and crews, that Bangor Marina adopted the dapper black guillemot as its logo. Across the road is the Guillemot Café. By now the sailors and the people of Bangor know these birds well and understand the need for their conservation. Sadly, Julian died in 2017 after thirty-two years studying black guillemots (reported in seventeen scientific papers and popular articles), but his legacy lives on, as he pioneered the methods of conservation for this species and ensured its future here. Shane Wolsey says, 'Julian's study and conservation of black guillemots at Bangor, County Down, exemplified his personal belief in the value of long-term studies and environmental monitoring that help us understand the natural environment. If only we could have more people with Julian's foresight and commitment.'

With time to spare, I walked a short distance to the east to visit the village of Groomsport. This was partly because my grandparents lived here a century earlier and I wanted to find out more about this branch of my family. I had hoped to check the local church records, but these were not available. Instead, I walked down to the small harbour nearby. In the 1840s this port was home to a small fishing fleet of nearly twenty vessels and eighty fishermen. There was probably a big seasonal influx of Scottish and Isle of Man fishing boats following the herring shoals down the Irish Sea. The villagers were mainly employed in farming, fishing and loom weaving, with women finding work in linen embroidery, locally known as 'sprigging'. As early as the 17th century, a line of small cottages called Cockle Row was built here, perpendicular to the sea, to protect the occupants from the strong north wind off Belfast Lough. My attention was drawn by the loud calls of seabirds on Cockle Island, a small island in the harbour that is particularly important for breeding terns. Between 150 and 200 pairs of arctic tern breed there, with a smaller number of common terns and a large colony of up to 500 pairs of sandwich terns. I went into one of the cottages on the harbour which has been set up as a Seabird Centre by the British Trust for Ornithology. Live images from several cameras on the island are beamed to screens in the Centre which give a unique opportunity to study the family life of the terns.

Leaving the safety of Bangor Harbour, we sailed east to the sound between Donaghadee and the Copeland Islands, through which the tide accelerates. The smaller two of the three islands are known as Lighthouse Island and Mew Island, but this is confusing at first, as the modern lighthouse is located on the latter. The original stone lighthouse was built on the former island in 1796 but this was demolished and a new lighthouse tower opened on Mew Island in 1884.[9] Praeger wrote:

> I remember as a boy charging boldly down on the Copelands amid swirling tides and wreaths of fog; finding Mew Island more by accident

> than by design, and visiting its wonderful tern colonies; and how thankful we were when tide-rips and mist were safely left behind.(BS)

The old lighthouse buildings now house the Copeland Bird Observatory, established in 1954 and one of only two continuously manned bird migration watch points on the island of Ireland.[10] Lighthouse Island has a cliff on the east side and fairly gentle slopes elsewhere leading to a rocky shoreline. For most of the visiting season, the island is covered by lush vegetation, including elder scrub, bracken, Himalayan balsam, nettles, bluebells, narcissi and sea campion. The observatory, which is run by volunteers, is the only place in Northern Ireland where rare songbirds are recorded with any regularity. I landed there some years ago and walked around the island taking a look at the big walk-in wire mesh cages, named Heligoland traps after the site in Germany where they were first used. Here, under licence, some migrant birds become trapped. Visitors can participate in the trapping, ringing and release of the birds and get a chance to see them at close quarters. The island also has a colony of almost 5,000 Manx shearwaters nesting in thousands of burrows below ground level. These are very long-lived birds (the oldest recorded at 55 years of age) and some marked individuals have been recovered off the coast of South America in winter. I had already seen streams of these birds flying north from their feeding grounds. Apart from the birds, the observatory is available as an ecological study centre, and is a valuable educational resource for students. There is accommodation for overnight visitors in the converted ruins of the old lighthouse.

Strangford Lough

The eastern coast of the Ards Peninsula is a relatively low rocky shore with many offshore rocks and shallows, a graveyard for sailing ships lacking accurate charts. About two-thirds of the way along its length is the tower of the old South Rocks light, completed in 1797 and still in good order. However, because it was really in the wrong place, the light was removed from the tower in 1877 and was replaced by a light ship two miles further out. In turn, this has now been replaced by a buoy. Rounding Ballyquinten Point, I was anticipating a rough ride across the mouth of Strangford Lough as the falling tide met the southerly winds blowing up the Irish Sea. But all was well, as the tide had not yet turned and was still flowing strongly into the lough. Strangford is effectively an inland sea over 30 kilometres from end to end, and with about 150 square kilometres of water surface. Twice each day powerful tidal currents flow in and out through the narrow entrance to this huge body of water. Bob Brown, formerly Head Warden for the National Trust here, wrote a wonderful description of the tide entering the lough:

> In the Narrows, running between Strangford and Portaferry, cut deep into the contorted Silurian bedrocks, a massive surge of water flows with every turn of the tide. A restless and uneasy combination of currents, whirlpools, upwellings and powerful back-eddies, it streams along, relentlessly tugging boats off course, carrying seals and dead branches in its flow; a powerful reminder that, in spite of being so land-locked, the Lough is very much part of the Irish Sea.[11]

I have passed through this remarkable channel several times under both sail and motor. One of the most extraordinary features here is a whirlpool called the Routen Wheel, where an upwelling of fast-moving water surges out over the surrounding sea, creating a smooth surface but swirling currents underwater that grab at the rudder and swing the boat off its course. The regular car ferry across the Narrows has to make a wide curve to compensate for the tidal surge, which reaches speeds of over seven knots (about thirteen kilometres per hour). Bob Brown told me that he and a companion once dived to the bottom of the Routen Wheel – in slack water, of course. They were very impressed by the seabed here, with massive boulders or rock outcrops and patches of very clean, coarse sand, almost like weathered coral. The rocks were smothered in millions of tiny hydroids, filter-feeding on plankton in the strong current. Slack water only lasts a few minutes, and eventually the divers were swept away, surfacing about a mile south to be collected by their dive boat.

Once safely inside the lough our relief was palpable as the currents spread out and slowed down over shallower waters. The sight that greeted me was a confusing contortion of 240 kilometres of shoreline dotted with about 120 islands and many more intertidal rocks, reefs and pladdies (a local name for shallow areas that dry at low tide). Sam Hanna Bell gave this description in his classic novel *December Bride*:

> Impeded by hundreds of islands, the waves never mounted to the fury of those of the sea, the menace lying in the currents that raced through the passages between island and island. The punt was now crawling across such a passage and approaching the deep channel where a swift racing band of water, broken and wind-blown, raised itself like the ruff of an angry dog. Then, as they neared mid-channel, Sarah felt a sudden exhilaration and a surge of kinship and love for her three companions. Her fear was subdued and lost in this feeling of kinship, of nearness to men who recognised the danger, accepted it, and were battling with it.[12]

The lough also has a wide range of marine habitats, from rocky shores with fast-flowing water at the mouth to huge tidal mudflats at the north end. Most of the islands are made of sand, shingle or gravel dumped here by the last glaciation 15,000 years ago.

As an outdoor laboratory the lough is unsurpassed, as it contains representatives of the majority of inshore subtidal habitats found in Britain and Ireland. Equally, the marine life of the lough is rich and varied, with thousands of species, many of them very beautiful. Some twenty-eight of Strangford's species are unknown elsewhere on the Ulster coast. In Portaferry on the eastern shore of the Narrows, I called in to visit the Marine Biology Station run by Queen's University Belfast. I had been here in the 1970s and was pleased to see that it was still being used by generations of biology students, while many other marine laboratories in Britain and Ireland have closed.

One of the rarest and most important habitats lies in deep water not far inside the Narrows. It has been created by large horse mussels that grow on the seabed here, creating a complex structure that is home to a bewildering range of animals, something like a coral reef community in tropical waters. A scuba diver as well as a respected conservationist, Bob Brown says 'in some areas it was impossible to see the bottom because of the tangled carpet of animal life'. Dense beds of mussels, forests of brittle stars, meadows of anemones have all attracted his attention over the years. Despite being a marine nature reserve, scallop dredging here destroyed large swathes of seabed habitat dominated by the biologically diverse horse mussel community. Bob says, 'destruction of the habitat of a species important to a commercial fishery must surely represent one of the most short-sighted actions to be undertaken by any industry. I still steam up when I think about it.'

Even the muddy areas at the north of the lough are packed with armies of burrowing worms and shellfish. The lough is a nursery area for a wide variety of fish and its birdlife is exceptional by any standards. In early autumn the vast majority of the world's population of light-bellied brent geese fly into the northern part of Strangford from their arctic breeding grounds to feast on the abundant seagrass beds. Large, noisy colonies of five species of terns nest on the numerous islands in summer. Praeger wrote:

> Memories come of low, gravelly islands in Strangford Lough, in June alive with nesting birds. One sailed from islet to islet, to be met at each by the shrieks of a white cloud of terns, the clamour of gulls, the sharp cry of the oyster-catcher. The fringe of grass-wrack that marks

> storm-level would be so thickly sown with eggs that one had to pick one's steps ... and all around were the waveless waters of the land-locked lough, reflecting the fertile undulating fields of Down.(BS)

While I was a nature reserve warden in Northern Ireland, I became interested in the group of harbour seals that hauled out on the sand banks near my house. I could hear their plaintive cries on the sea wind and I found their behaviour fascinating. The harbour seal is the smaller of the two species breeding in Irish waters and, with colleagues, we found one of the biggest colonies of these seals at the time in Ireland in the sheltered waters of Strangford Lough. The numbers here reached 800 in the late 1980s. Then a disease hit them and the population collapsed.[13] The pups are born in June and July, and they stay close to their mothers for weeks after this, sometimes suckling in the water, even riding on their mothers' backs. The behaviour changes during the year, with increasing time each day spent hauled out of the water while the seals are moulting their thick coats in late summer. Grey seals also breed in Strangford Lough and, while their numbers collapsed too in the early years of this century, they are now thankfully increasing again.

The human story of Strangford is equally fascinating, from the days of the Viking ships to the present popularity of these sheltered waters for sailing and other water sports. At Nendrum on Mahee Island an early Christian monastery contained a unique form of renewable power. Excavations on the shoreline here found the remains of a horizontal mill that was built into a stone embankment designed to trap the waters from the lough. As the tide fell outside, water cascaded over the mill wheel, providing endless energy for grinding corn. Timber within the structure has been dated to 620 AD, making it one of Europe's earliest examples of the use of tidal power.[14]

South Down Coast

Leaving Strangford Lough, we sailed past the headland of Killard Point, Gun's Island and Sheepland, scene of the *Amitie* shipwreck of 1797. This was an attempt by France to assist the United Irishmen by landing a consignment of guns. There may be a link between this and the name Gun's Island, although the sites are about a mile apart.[15] After a long day at sea, we were happy to reach the sheltered harbour of Ardglass. The marina for visiting yachts here consists of just three pontoons and, despite its small size, it is among the best-run facilities of its kind in the country. Fred Curran and his dog have been managing this small marina since the 1990s. It is tucked into a corner of the harbour that also holds one of Northern Ireland's main fishing ports. Sleeping on a yacht on the marina can be a little difficult with the continual background rumble of the engines of a dozen fishing trawlers heading out at five o'clock in the morning. However, there is some compensation in the natural sounds of grey seals yawning, herons screeching, whimbrels and curlews whistling. Praeger described Ardglass in the 1930s as 'a busy centre of the herring fishery and, being a breezy and picturesque place'.(WW) At certain times of year, in the late 19th century, there would have been hundreds of wooden sailing boats tied up in the harbour here as bands of migrant herring fishermen from Scotland, England and the Isle of Man chased the huge shoals of 'silver darlings' in the Irish Sea.

I always take extra care leaving Ardglass through the narrow channel to the Irish Sea, as a friend of mine ran his boat over an unmarked rope from a lobster pot here and the resulting tangle around the propeller nearly caused a shipwreck. Just to the south-west is the distinctive yellow-and-black lighthouse on St John's Point. The playwright Brendan Behan was an assistant here about 1950, and the principal keeper, Mr Blakely, had to get him sacked for slovenly behaviour, absence without leave, foul language and worse.

On this coast, the southerly flow of the tide entering the Irish Sea past Antrim slows down as it meets the northbound flow of tide from St George's Channel to the south of Wexford. Running from St John's Point to the Isle of Man is the Irish Sea Front, where warmer surface layers flow over deep, colder waters, creating a circular gyre in some seasons. The other large front off Northern Ireland runs from Malin Head in north Donegal to Islay in Scotland, and across which there is a marked temperature change. This temperature difference in the seawater affects marine invertebrates such as the thick-shelled topshell, which is abundant to the west of Malin Head but scarce further east until it occurs again to the south of the Irish Sea Front. These fronts are places where the upwelling of plankton attracts giant basking sharks to feed. There are particular concentrations of these impressive fish around Malin Head and near the Isle of Man.[16]

West of St John's Point, the coast opens out into the wide, shallow waters of Dundrum Bay dominated by the rounded slopes of the Mourne Mountains. Here the passenger ship *Great Britain*, designed by the famous engineer Brunel, was bound for New York in 1846 when she went aground near Tyrella, a few miles to the west of St John's Point, due to a navigational error. At Brunel's suggestion a massive breakwater was erected to protect her and, almost a year later, she was refloated and recommenced her career. She is now a floating museum in Bristol.

Dundrum Bay

From the summit of Slieve Donard, the Mountains of Mourne 'sweep down to the sea' and the long sand dune system at Dundrum Bay where I began my trade as a field naturalist. As I related in the introduction to this book, I arrived on the dunes at Murlough Nature Reserve quite unexpectedly in the early 1970s, fresh out of university. It was, as now, a stunningly beautiful environment for a young graduate to put into practice some of the natural science that had been absorbed in lectures and practical classes. Originally owned by the Marquis of Downshire, the estate had been acquired in the 1960s by the National Trust for Northern Ireland. I was the second warden to be appointed to the staff and it was to influence the rest of my life.

The dunes form a long spit that stretches out parallel to the coast from the town of Newcastle and is separated from the village of Dundrum by a small estuary and a narrow channel through which the tide rushes twice each day. During the Second World War, these sand dunes were used for military training, and occasional remnants of concrete and steel still remain from those days. The beach here must have also been important for training, as it was similar to those beaches invaded in the Normandy landings of 1944. Jo Whatmough, the first warden at Murlough, now thinks that rabbit farming or 'warrening' was of immense importance in the past land-use history of the dunes. It had probably started on the dunes by the 13th century, following the construction of a Norman Castle nearby, and descriptions of the dunes as the Greater Rabbit Warren in the 18th and 19th centuries refer to the area as a 'sandy waste'.

However, during the 1950s, the disease

myxomatosis decimated Ireland's rabbit population, and the natural dune vegetation began to recover. When the National Trust took over in the 1960s the priority was to reduce erosion caused by increasing recreational use of the dunes and beach. These ideas have changed over the decades, and there is now a view that instability is a natural feature in sand dunes, and certain scarce species such as orchids and some winter annual plants depend on the absence of competition from more vigorous vegetation.

One of the unique features of the Murlough dunes is the full sequence of habitats, from shoreline and foredunes through fixed dunes to dune heath and woodland on the oldest and most stable part of the system. Dry, stable sand dunes change over thousands of years, becoming more acidic as the lime is leached from the soil by rain. The first stage is a dense growth of bracken and burnet rose, which is replaced by heathland as the acidity increases. Here the vegetation is dominated by ling and bell heather, with typical mountainy plants such as tormentil, heath bedstraw and a selection of acidic mosses and lichens.[17]

However, this natural succession has been somewhat disrupted by an introduced shrub, sea buckthorn, which was planted here over a century ago in an attempt to stabilise mobile sand. In the 1970s, control and removal of the dense thickets of this plant were among the main jobs of the wardens. Jo Whatmough started work here as a National Trust warden in 1967. Now retired, she still gives volunteer help and advice on the reserve. She remembers that, when she started work first, she had a largely free hand to manage this important site for conservation. It was a case of learning by doing, as she developed new techniques such as the use of wooden boardwalks for visitors to reach the beach without causing further damage to dune vegetation. The path network is now estimated to carry over 250,000 visitors per year.

I was curious to find out how the priorities have changed over the near half-century since I started work here. The present National Trust warden, Patrick Lynch, took me on a tour of the dunes in early summer. One striking change since my time here is the openness of the dune vegetation. Large areas have been cleared of invasive sea buckthorn, and the sycamore woodland that followed it has also been felled and removed. Heather-dominated vegetation has become more widespread and rabbits were even reintroduced to try to restore some dynamism to the dune vegetation. A herd of Exmoor ponies was brought here in the 1990s from a local rare-breeds farm. They have had dramatic effects in reducing the height of the grassy vegetation, controlling the spread of bracken and preventing gorse scrub from overrunning the dunes. Jo explained: 'This breed was chosen as they were considered to be the best for grazing of heathland in particular. They tend not to damage the heather but they eat the young tips of gorse. This seemed to be important for them during the long summer drought of 2018.'

However, Patrick's experience in managing this beautiful coastal site is much more strictly controlled now by the fact that the reserve is designated as a European Special Area of Conservation. This has brought more regulations on which habitats must be protected and the ways in which they can be managed. Together we walked across the dunes to the beach with its iconic view of the Mourne Mountains. The sea has brought dramatic changes to the front line of dunes above the beach, with massive erosion causing slumping and then revegetation of the sand by marram grass. But, as Patrick explained, this has freed up the sand to move naturally between beach and dunes, allowing a much more natural evolution of the system.

I was impressed by how nature conservation policies and priorities can change over time. The value of practical management only becomes evident in the long term, as nature adapts to new situations such as natural pressures from the sea and from grazing animals. The natural cycles of erosion and accretion often balance each other if they are only given time and space to do so. However, after half a century of dedication to this project, Jo Whatmough is more pessimistic about the future of the natural habitats here. 'I wish I could claim that the conservation of the wonderful habitats of dune grasslands as well as dune heath, had been successful. Although habitats on the site are now seen as of European importance, many of those that were here when I arrived in 1967 are much reduced, some even no longer present.' With the benefit of hindsight, she can see that there are much greater forces at work here, including the effects of long-term climate change.

Mourne Country

The Mourne Coast from Newcastle to Kilkeel includes the mouth of the Bloody Bridge River and the Brandy

Pad leading up and across the saddle near Slieve Donard. The car park here is a great vantage point from which to see passing harbour porpoises. The river gullies that flow down from the Mourne Mountains have fine examples of natural vegetation that is totally uncultivated. The intimate little harbour of Annalong has a fine restored water mill, from the 19th century or earlier. Here there is a beautiful species called the oysterplant that I once found on the shingle beach nearby. This is a rare northern flower that survives only in a few exposed northern locations in Ireland. It grows alongside the yellow horned-poppy, which is here at the northern edge of its range.

This area was the subject of a classic book entitled *Mourne Country* by Estyn Evans, Professor of Geography at Queen's University Belfast. Evans was a Welshman and friend of Praeger, to whom he dedicated his book. He described this isolated peninsula between the mountains and the sea as the 'Kingdom of Mourne', with its special folk traditions, fishing and farming practices.

Whether seen from the sea, the air or from a distant viewpoint to the north, the Mountains

> of Mourne are a remarkably compact and clean-cut range. From Newcastle to Rostrevor, where they tumble into the sea in Dundrum Bay and Carlingford Lough the distance in a straight line is 14 miles. You can walk this in a summer's day and, if you keep to the peaks in 'the back side', away from the sea, you need never drop below 1,200 feet on the way.[18]

Indeed, on a clear day I have often seen the distinctive outline of the Mournes when standing on the summit of Howth Head some 125 kilometres to the south.

Just past the fishing port of Kilkeel is the navigation mark known as the Hellyhunter that marks the entrance to the shipping channel for Carlingford Lough. I sometimes pause here and reel out a fishing line as it is a good place for mackerel in the summer. There is nothing more delicious than a couple of fresh fish sizzling on the pan as my yacht enters the narrow waters at the mouth of the lough. This is a classic fjord, one of just two in the island of Ireland, with a shallow rocky entrance and deep waters inside. It was described by Gerard Boate in his *Natural History of*

Ireland, first published in French in 1666.

> This haven is some three or four miles long, and nigh of the same breadth, being everywhere very deep, so as the biggest ships may come there to anchor; and so inviron'd with high land and mountains on all sides, that the ships do lye defended off all winds; so that this would be one of the best havens of the world, if it were not for the difficulty and danger of the entrance, the mouth being full of rocks, both blind ones and others, betwixt which the passages are very narrow; whereby it cometh that this harbour is very little frequented by any large ships.

Despite this warning, the shipping channel today is well marked with red and green buoys where it passes between the villages of Greencastle to the north and Greenore to the south. Carlingford Lough is divided down the middle by the international border between north and south. Many years ago, while censusing seals here with a team of biologists, our inflatable boats were 'buzzed' from the sky by a British Army helicopter as this was during the Troubles, when there was a heavy military presence along the border. Luckily, we were able to communicate with the aircrew by VHF radio and we stated our innocent intentions.

Further into the lough is the large harbour of Carlingford nestled beneath the Cooley Mountains and dominated by an impressive castle. Built in the 12th century by the Norman knight Hugh de Lacy, the building is known as King John's Castle to commemorate a visit by the monarch in 1210.

On the north side of Carlingford Lough is Mill Bay. It looks like a typical mudflat today, but during the 19th century this was covered with seaweed. This was due to large-scale cultivation of weed for the manufacture of kelp, which was sold and also used as fertiliser on the local farms. Local people, who claimed rights to sections of shoreline, laid out rows of large rocks in lines across the low tide area to which the brown seaweed attached in abundance. At one stage there were over a thousand of these seaweed beds covering an area about the size of the nearby town of Greencastle. In the depression between the two world wars, this local industry became neglected.

Sailing south from Carlingford towards Dublin, my friend John and I were not alone. A school of up to a dozen common dolphins joined the yacht, playing with each other in the bow wave from the boat like children released from a house to race around the garden. In a flat sparkling sea, they sped along beside the boat for over an hour, leaping from the water, slapping the surface with their tail flukes and even rubbing their bellies together as they swam at high speed. Common dolphins can swim at speeds up to twelve kilometres per hour, occasionally powering ahead of the boat like torpedos, then circling around and surprising us as they again approached from behind. This joyful spectacle continued almost to the coast of Dublin.

Rockabill

Lying off the seaside town of Skerries, the dramatic island of Rockabill with its iconic black-and-white lighthouse is largely deserted in winter but has a huge population of nesting terns in summer. Terns are among the seabirds that undertake long annual migrations to west Africa and back each year. Sandwich terns are the first to arrive in late March, with the bulk of the other species following in April and May.[19]

In the 1980s the Commissioners of Irish Lights made a decision to automate the majority of the lighthouses around the Irish coast, installing computer controls instead of the lighthouse keepers, many of whom had followed their fathers and grandfathers into the profession. One of the nearest lighthouses to Dublin is on Rockabill. Here, the dramatic lighthouse tower and buildings are perched on top of the largest of two small rocks, surrounded by miles of unbroken sea. I was working at that time in BirdWatch Ireland and made several visits to Rockabill where it was exciting to see so many terns again after destruction by the sea of the big colony at Tern Island in Wexford Harbour. Common terns were the most abundant nesting species then, but of special interest was the roseate tern, a declining species right across its range in Europe. Here the birds chose to nest in among the tall stands of tree mallow that had colonised the old lighthouse gardens. For years, the lighthouse keepers had acted as unofficial wardens of the tern colony, ensuring that there was no unauthorised disturbance at critical times of the nesting season. Automation of the site presented both a challenge and an opportunity for conservation of the terns.

Without the keepers present anyone could land on the island and cause the terns to desert but, worse, there was always a threat of either deliberate or accidental introduction of ground predators such as cats or rats which would cause devastation among these ground-nesting birds. However, the vacating of the island buildings offered a potential base for summer wardens. So, with the agreement of the Commissioners, we recruited the first in a series of enthusiastic ornithologists to spend several months each year managing the tern colony. Living for weeks at a time in fairly basic Victorian buildings, on an isolated rock in the Irish Sea surrounded by the screaming of thousands of seabirds, required both dedication and resilience. Fortunately, the first of the wardens in 1989 was Liam Ryan, who came from a Wexford family of lighthouse keepers – so he knew about isolated living conditions. Liam has been followed over the intervening thirty years by dozens of equally committed wardens who have ensured the protection and growth of the tern colony.

Approaching Rockabill thirty years later by yacht, I sailed around the shore of the island which looked like a huge beehive with clouds of white seabirds in the air all over the conical rocks. I dropped anchor in a sheltered creek between the two rocks and rowed ashore in my small dinghy to meet the wardens. The first job for the wardens each spring is the clearing of the previous year's growth of tree mallow to create the space for thousands of tern nests. Then there is the deployment of hundreds of wooden nestboxes as shelters for the birds to lay their eggs out of sight of the larger gulls which would dive in for an easy meal if the nests were exposed. Many of the nestboxes were made by the carpentry classes of a local school in Balbriggan and some of these are placed in rows on rocky terraces like a large housing development. Finally, the wardens have to census the colony, monitor the number of eggs and chicks and place numbered rings on the legs of a sample of birds. This allows them to be followed in later years as the younger birds return to nest on the island – or spread out to other colonies. This bird is a global traveller and there have been a number of spectacular recoveries of Rockabill terns migrating to other continents including west Africa, Brazil and eastern USA.

The availability of suitable fish stocks in the waters surrounding this tern breeding colony is vital to its success. The roseate tern feeds mainly on sand eels, juvenile herrings and sprats, which are also key food sources for many other seabirds. Unfortunately, there is currently no scientific monitoring of these fish stocks and hence no early warning system for any collapse in the prey populations. BirdWatch Ireland researchers have recently undertaken high-speed boat trips following the terns from Rockabill out into the Irish Sea to observe where and on what fish they are feeding.

The results of all this work, which is supported by the National Parks and Wildlife Service, have been dramatic. At the start of the project in 1989 there were less than 200 pairs of roseate terns on Rockabill. Over three decades later the colony has increased rapidly and, in 2021, contained the nests of 1,704 pairs of roseate terns and 1,656 pairs of common terns – the largest single colony of the former species in Europe.[20] With high breeding success in most years, roseate terns hatched on Rockabill have moved to nest in several other declining colonies around Ireland and Britain, thus spreading the population over a wider range. This is a highly successful conservation project which has developed new techniques that have been copied in a number of other locations.

Brian Burke worked for three successive summers as a warden on Rockabill. He says, 'the more you watch a species, the more you learn about the "character", including its motivations and the struggles it has gone through to get where it is today. It was a real privilege being able to spend every single day of the summer watching the terns go through the highs and lows of the breeding season, having travelled all the way from Africa in the hope of finding enough fish to feed their chicks. Even when the work for the day was done, we would stay outside and watch everything happening around us.'

Brian believes that the success of the Rockabill tern conservation project over the last three decades has been staggering. 'Looking forward, we hope that the success of Rockabill will contribute to the establishment of roseate tern colonies elsewhere in the Irish Sea and north-west Europe to help provide some population resilience for the looming challenges of climate change and declining fish stocks in this part of their range.'

Lambay

I have been on the island of Lambay on numerous occasions, including a visit in the 1980s with then President of Ireland, Patrick Hillery, to meet the owner, Lord Revelstoke. This is the largest island on the east coast and it remains a privately owned wildlife sanctuary. Praeger related how in 1904 the new owner of the island, the banker Cecil Baring, came to the National Library seeking information concerning his new possession. 'He told me how in Munich in 1903 he and his wife saw in *The Field* an advertisement "Irish island for sale", and how they promptly bought it and set about making habitable its old castle. I think the Barings would have willingly spent their lives on the island, for they became intensely interested in everything it contained.'(WW) Baring was keen to 'enhance' nature on the island, so he introduced a herd of fallow deer and, curiously, a group of wallabies from Australia. The descendants of both species are still present today.

With common interests in natural history, Praeger and Baring quickly became friends and the former wrote how he and his wife spent 'many a delightful holiday' on the island as guests of the Barings. Praeger, who was deeply interested in island biogeography, the unique assemblage of species on an isolated area of land, relished this opportunity and wasted no time. 'Shortly after he took possession of the island, I suggested to him that a detailed study of its natural productions –

animal, vegetable and mineral – would be interesting, and might have important scientific results. He accepted the suggestion at once.' Praeger immediately set about organising a team and, during 1905 and 1906, twenty different naturalists stayed on Lambay and, in his own words they 'ransacked the island from end to end'. The result was a series of papers, published in the *Irish Naturalist* on the natural history of Lambay. Some of the animal groups were poorly studied at that time so it does not seem surprising today that the surveys resulted in five new species to science – three worms, a mite and a bristletail – and between eighty and ninety animals and plants that were previously unrecorded in Ireland.[21]

Prior to the Barings' ownership, the botanist Henry Hart spent four days on the island in the summer of 1881, and four further days in the spring of 1882, accompanied by the naturalist Richard Barrington. Hart undertook a comprehensive survey of the plants on the island, and one of his more striking discoveries was a colony of over half a hectare of a very rare plant, the adder's-tongue fern, with a 'dozen plants in a 6-inch square sod'.[22] During this period the people living on the island had left and a herd of fallow deer was introduced by the owner to develop a shooting estate here. Previously cultivated areas became overgrown and by the time of both Hart's and Praeger's visits, Lambay was probably experiencing a remarkable reversion, with a resurgence of 'native' vegetation such as heathland covering formerly cultivated ground.

A century later, Matthew Jebb, Director of the National Botanic Gardens, undertook a repeat survey of the plant communities on Lambay. His findings present a stark contrast to those of both Hart and Praeger. Seventy-eight plant species, formerly recorded for the island, have not been seen in recent years. A number of alien (or garden) plants have been introduced in their place. Heather areas, once widespread, have now disappeared completely. This shows clearly how land use can have significant impacts on the habitats and species in an area over time.[23]

In 2005, I made a number of visits to the island with another group of biologists, this time to census the large colony of grey seals that breed there. The main groups haul out on the cliff-bound eastern side and the easiest way to see them is from a small inflatable boat that can get in close to the rocks without risk of damage. The seal pups are born on some steep shingle beaches at the head of narrow gullies and especially in the numerous caves where they are out of sight and protected from storms. The white-coated pups are unable to swim for the first three weeks of life, so they lie in these sheltered places growing fat on their mother's milk alone. Praeger was also fascinated by the grey seals. 'Once I stalked a group of them there on a flat tidal rock until I lay among them and could count every bristle and hair on their mastiff-like faces; they watched me closely, but never stirred. And we had delightful adventures with white baby seals, quite devoid of fear, which bumped against the boat and let us stroke them while a watchful mother swam continuously and silently round.'(WW) Such interactions with wild animals are not encouraged today.

On numerous trips since then I have sailed around the island admiring and photographing the impressive seabird colonies here – the largest on the east coast of Ireland.[24] Here the populations of common guillemots, razorbills, kittiwakes, herring gulls and shags are of special importance. By 2018 the huge guillemot colony had reached almost 60,000 individuals.[25] Some studies of the shags have been undertaken here by Steve Newton and colleagues from BirdWatch Ireland. By attaching GPS devices to the birds they were able to demonstrate that the majority flew in a direct line to feed on the Kish Bank off Dublin Bay. Here the shallow waters provide ideal feeding grounds for the shoals of small fish that the shags bring back to chicks in the nests. Worryingly, the shag population of Lambay has fallen by almost 60 per cent in the last twenty years, suggesting that they are experiencing some difficulty finding enough fish.[26] In contrast, a colony of gannets has become established on Lambay in recent years, and this had grown to reach 926 nests in 2015.[27]

Lambay was settled in ancient times and was important to Neolithic people as a source of the rock known as porphyry, which has distinctive white crystals.[28] This hard rock was valuable for making stone axes, which could then be used to fell trees. The quarry site on Lambay is unusual for being the only Neolithic stone axe quarry with evidence for all stages of production, from quarrying to final polishing. Lambay was also one of the first landfalls in the Viking invasion of Ireland. In 795 AD the raiders sailed down the Irish Sea, their target being a small monastic settlement on the island. The monastery, built in 530 AD by an Irish scholar named Colmcille, was home to a small band of monks and scribes. A church, monastery and several small, corbelled stone huts and simple shelters were built here. The monks possessed little. The only items of value were the books, religious artefacts and holy relics that adorned the church and monastery walls. The Vikings used Lambay as a base for raids on the more substantial settlement of Dublin (*Dubh linn*) and eventually established the first Viking town in Ireland here.

As I sailed past the eastern promontory of Lambay I could see the cliff-bound cove to its south known as Tayleur Bay. Beneath the strong tidal stream here lies the wreck of a ship named the *Tayleur*, which sank in 1854 with the loss of 360 lives. It was one of the largest iron sailing ships in Europe at that time and, like the later doomed passenger ship *Titanic*, was the pride of the White Star Line. Launched the previous year in Liverpool, she had embarked with 650 passengers en route to Australia and a new life in the British colonies of that century. Unfortunately, a combination of bad weather, poor navigation equipment and incompetence of the crew led the ship to strike Lambay's cliffs broadside on. The deck was said to be so close to the rocks that some men were able to jump ashore, but the majority drowned when the ship heeled over and sank.[29] With this sobering thought in mind, I was careful to give the island and its fearsome reputation a wide berth on this occasion.

Ireland's Eye

The craggy island of Ireland's Eye stands prominently across the mouth of the harbour at Howth. It is hard to believe that this wild maritime landscape, with its spectacular cliffs, is just a few kilometres from Dublin city centre. This is an important site for breeding seabirds and also supports a number of protected habitats, rare plant species and several important cultural heritage features, such as a Martello tower and the ruined St Nessan's church. The island is uninhabited at present but there is some evidence that it was previously farmed. The island has no built harbour and is generally accessed by small boats from Howth Harbour in good weather only. It is a popular destination for day-trippers and is visited by thousands of people each year between April and September.

Until recently, Ireland's Eye was owned by the Gaisford-St Lawrence family, whose impressive home for many centuries has been Howth Castle and Demesne, overlooking the island. The lands in Howth, which included Ireland's Eye, remained in the ownership of this family for over 800 years. Praeger described Howth as 'a delightful old-world place. Even round its rock-bound margin houses were few and one could wander at will along its grassy slopes and over its broad heathery top.'(WW) Today, Howth has become a sprawling suburb of Dublin and is packed at weekends with day-trippers from the city.

With increasing numbers of visitors to the island, Fingal County Council took a special interest in 2016 in the context of the newly designated Dublin Bay Biosphere. It commissioned several key surveys from experts to define exactly the importance of the island for plants and habitats, seabirds and cultural heritage. Amid reports of increasing disturbance of the important seabird colonies by visitors and a few cases where gulls had attacked people to protect their nests, a network of paths was planned to improve access for visitors while simultaneously leading them away from the main gull colonies. This work was initiated in 2017 and completed in a day using a team of workers with hand-held strimmers. The presence of a number of invasive plant species was identified as a problem, as they have the potential to compete with and threaten some native plant communities. A start has been made in controlling these by physical removal.

The same year I was commissioned to prepare a comprehensive management plan for the whole island. Part of this work involved several visitor surveys during the summer. My son, Tim, and I travelled to the island, clambering out of the small boat onto the rocks beneath the impressive Martello tower. We climbed the steep slopes to the summit of the island, from where we could see all the paths, count the number of visitors landing during the day and map their distribution on the island. Surprisingly, the majority stayed around the landing place and did not explore the rest of the island at all or simply wandered down to the sandy beach for a picnic. The mown paths had clearly been a success, as most of the visitors remained on these and did not stray into the gull colonies.

We planned a series of low-impact signs at various strategic points to guide the visitors to the places of interest. The large and impressive cliffs at the eastern end of the island are a big attraction for birdwatchers. Here gannets nest on all the steepest slopes with crowded lines of guillemots and kittiwakes keeping up a continuous racket right through the day. On the top of the cliffs there is a large colony of cormorants and these are vulnerable to disturbance especially when the flightless chicks are sitting in the big nests. We recorded several dogs off their leads causing the adult birds to fly and we had to recommend that dogs should not be allowed on the island at all during the bird breeding season. Fingal County Council is now planning to improve the landings at the west end of the island to allow safe berthing of the ferries and disembarking of passengers.

Baldoyle Bay

Just to the west of Ireland's Eye, and north of the Sutton promontory which links Howth to the mainland, lies one of the three estuaries of Fingal. Baldoyle Bay is almost surrounded by land except for a narrow entrance between Sutton and Portmarnock dunes. Most of the sand dunes are occupied by a large golf links and it was on the practice greens here that I found myself a few years ago watching brent geese through a telescope. The study, commissioned by Fingal County Council, was to monitor waterbirds on the lands around the estuary to ensure that enough open space was retained to accommodate them in this fast-developing suburb of Dublin.

Brent geese start each winter feeding in the Dublin estuaries but, when the seagrasses that they favour are exhausted, they move to foraging on coastal parks, sports pitches and golf courses. It is amazing to see a flock of over a thousand wild geese, recently arrived from their arctic breeding grounds, contentedly grazing in a park surrounded by people playing football, dogs being exercised and schoolchildren on their lunch breaks. Although constantly vigilant, the geese have become habituated to this human environment which gives them the essential forage that they need to survive till the return migration in the spring. What made this study so interesting was that a sample of the geese carried individually numbered rings on their legs, each with a unique colour-letter combination that allows identification, just like vehicle registration plates. This ringing programme is coordinated by the Irish Brent Goose Research Group. By recording a large number of these 'colour-ringed' geese I was able to map their movements between the grasslands and the estuary on a daily basis, and thus uncover the order of their preference for different sites. Sometimes I would arrive at dawn, as a big red sun rose from the Irish Sea, and watch a large V-shaped flock of geese flying in from their night-time roost on the North Bull Island in nearby Dublin Bay.

Dublin Bay

Framed perfectly between the rocky headlands of Howth and Dalkey, Dublin Bay has been the cradle of the capital city since it was established as a permanent settlement by the Vikings in the 8th century. I grew up on its shores, as did generations of my ancestors. I learned to swim and sail and fish here, and I have a deep familiarity with its waters and its shorelines to this day. Dublin Bay is so rich in nature and history that it is hard to know where to begin. Perhaps the words of James Joyce, who set his famous novel *Ulysses* here, are the most lyrical. 'Stephen closed his eyes to hear his boots crush crackling wrack and shells. You are walking through it howsomever. I am, a stride at a time.' I have experienced the same crackling sensation while walking over thousands of empty razorshells cast up on the beach at Sandymount Strand.

Up to the late 19th century shellfish were regularly sold on the streets of Dublin city and up to seventy people, including many children, were engaged in collecting cockles on the wide strands at low tide. This was cold and difficult work that involved wading in shallow water to rake the shellfish from the top layers of sand. Records kept by coastguards show that between 80 and 104 tonnes of cockles were landed annually, and probably a lot more which went unrecorded.[30] Throughout the 18th and 19th centuries there were a number of typhoid epidemics in the city which were blamed on poor, unsanitary living conditions of the population. By 1904 this problem had also been linked to the gross pollution of Dublin Bay with untreated sewage and the consumption of contaminated shellfish. In 1890, five members of one family in Blackrock died from eating contaminated mussels. In the six years before the First World War, the landings of cockles declined dramatically and shellfish collecting for human consumption has been banned here ever since. The story of Molly Malone in the song 'Cockles and Mussels' is surprisingly close to the truth.[31]

The abundance of shellfish and other marine life in Dublin Bay is the main reason that it holds one of the largest concentrations of migrant shorebirds in Ireland. Huge flocks of brent geese, oystercatchers, godwits, dunlins and redshanks crowd into their high-tide roosts on Sandymount Strand and the North Bull Island. Since 2013, I have been fortunate to be involved in a large monitoring project on the Dublin Bay birds run by BirdWatch Ireland and funded by Dublin Port Company. We have had several exciting adventures, with a larger team of ornithologists, cannon-netting a sample of the birds to give them numbered leg rings and, in a few cases, satellite transmitters. Information gained from the subsequent observations and GPS fixes has added enormously to the information necessary for their conservation. For example, we now know that internationally important numbers of some species are present throughout the year, that some individuals return each year to feed on exactly the same stretch of shoreline and that in the summer most of our oystercatchers move to Scotland, Norway and Iceland to breed.

The long spit of Bull Island on the north side of the bay has one of the best-developed sand dune systems on the east coast and has a very large area of saltmarsh that the birds use as a safe refuge. The island itself is the result of engineering works undertaken at Dublin Port in the 1820s which created a retaining wall on the north side of the Liffey channel to aid navigation of ships in the shallow waters. In 1800, Captain William Bligh of the Royal Navy undertook a comprehensive depth survey of the entire bay. Bligh had previously become famous through his involvement in the mutiny on HMS *Bounty* in the Pacific Ocean and his extraordinary 4,000-mile voyage to safety in an open boat. Following Bligh's survey, the North Bull Wall was built from Clontarf to the mouth of the Liffey. This had the desired effect of scouring out the sand bar at the mouth of the river and this sediment gradually redistributed itself on the northern shores, quickly forming a new island that now measures about five kilometres long and one kilometre wide. Today it is a popular recreational resource used by thousands of people and their dogs. When I am surveying bird flocks through a telescope here, I am regularly asked by passers-by if I have seen any rare species. They are usually stunned to learn of the vast numbers of common species such as waders and gulls using the bay.

Dublin Port

I often sail my yacht past the iconic red Poolbeg

Lighthouse and into the mouth of the River Liffey, where enormous amounts of seawater and freshwater mix twice each day. The passage leads beneath the twin chimneys of the Poolbeg electricity generating station which are visible from most parts of Dublin. My destination is the sheltered marina run by Poolbeg Yacht and Boat Club, right in the centre of the port. My small boat is dwarfed by enormous cruise ships, cargo vessels and ferries in the river channel.

Black guillemots breed in many ports and harbours around the Irish coast, including Dublin Port, where up to fifty birds nest in a variety of structures, including holes in stonework and boarding ramps. With the modernisation of the quay walls many of these old cavities were replaced with concrete so Dublin Port Company provided a series of custom-made nestboxes, based on the experience of Julian Greenwood in Northern Ireland. These allow the birds to continue breeding close to human activity. I have been surveying these dapper black-and-white birds here annually over the last decade and I have always been impressed by how unconcerned they are by the shipping traffic and noise in the busiest port in the country. I stand on the deck of a launch, carefully scanning the quay walls all the way from the Poolbeg Power Station to O'Connell Bridge in the centre of the city. The birds poke their heads out of drainage holes in the older quays, sit in pairs on the top of the quay walls or on the water or even perch on the hulls of giant cargo vessels. They fly rapidly just above the water surface heading out into the bay to feed on small fish.

Terns have also been known to breed in Dublin Port since at least the 1940s. In 1995 the late Oscar Merne began a long-term study of the colony, which was then nesting on a disused mooring structure close to the ESB Poolbeg Power Station. He published an early analysis of his results from the period up to 2003.[32] From 2012 the monitoring work was carried on by Steve Newton with a team of colleagues from BirdWatch Ireland. The main colony continued to grow and the terns were recorded trying to nest on moored boats, on the lock gates of the Grand Canal and various other unsuitable places. In 2013, Dublin Port Company decided to help the terns by providing a floating pontoon to allow the colony to expand. Moored between the port and Clontarf, this platform had wooden planks around its edges and a covering of gravel to mimic a stony beach. The terns readily adopted this artificial site so, in 2015, a second larger pontoon was moored off the Great South Wall and this was also used for nesting by both common and arctic terns.

In the winter of 2016–17 the wooden part of the original mooring structure, now over a century old, began to collapse into the sea. So the ESB undertook to rebuild this with a custom-made tern-nesting platform, benefitting from the successful trials of Dublin Port. By 2020 the entire colony, now nesting on four completely artificial structures, numbered almost 600 pairs, with some of the terns originating from Rockabill off the north Dublin coast. A thriving colony of these threatened seabirds nesting successfully each year in Dublin Port, close to the busy shipping traffic, is a clear demonstration of the benefits of conservation management. Developing and operating port infrastructure inevitably impacts the natural environment. However, as Eamonn O'Reilly, Chief Executive of Dublin Port Company says, 'with a commitment to, at the very least, conserve the natural capital of the Port, there is no reason why such impacts cannot be positive.'

Dalkey Island

The southern arm of Dublin Bay stretches out beneath the hills of Killiney and Dalkey, ending in the rocky Dalkey Island. I love to sail with a fast-moving tide through the narrow sound between the island and the mainland, passing by the tiny stone pier at Coliemore Harbour. From the island side my yacht is watched by a group of grey seals resting on rocks close to the landing steps, and by a herd of wild goats grazing peacefully by the Martello tower which dominates the summit of the island. This is one of a string of twenty-six similar towers that stretched the full length of the Dublin coastline from Balbriggan to Bray, although only twenty-one of these still stand today. They were built in a very short time in 1804–05 when the British military authorities decided there was a good chance that Napoleon's forces would attempt a landing near Dublin to take what was then regarded as the 'second city of the Empire'. Eight of the towers had associated gun batteries; the one on the south end of Dalkey Island is especially impressive. Walking around the substantial battlements, built of huge cut-stone granite blocks, I tried to imagine a fleet of French ships, their square sails reflecting the morning sun, appearing around Bray

Head to the south, and the small garrison on the island called to action stations with orders shouted by their officers. But Napoleon's navy never did invade Ireland, and the towers were eventually decommissioned to remain empty for the next two centuries.[33]

On one of my many visits to Dalkey Island I met with the archaeologist Jason Bolton, who showed me several temporary settlement sites from ancient times. The most obvious signs are layers of empty seashells such as limpets well below the present ground surface but visible because waves have eroded away the sand at the back of the beach. These sites were excavated by David Liverage in the 1950s and have recently been reinterpreted by Barbara Leon. As well as the shell middens (or dumps) there was a collection of flint artefacts (flakes, scrapers and blades), hammer stones, arrowheads, stone axes, a grinding stone and primitive pottery. Postholes beneath these deposits indicate temporary shelters on the site. Other excavations on the island revealed bones of cattle, sheep, seals, fish and birds, as well a human burial which was dated to around 5,000 years before the present. Piecing all of this evidence together, I imagine a band of Stone Age people coming here repeatedly in summer to collect shellfish, seaweed, fish and perhaps to hunt birds and seals. The evidence for the structures suggests that the island was occupied at least for short periods, and the placing of a group of flint blades and flakes in a distinctive downward-facing position also suggests that it was used for occasional ritualistic ceremonies such as burials.[34] Mesolithic people were widespread in Ireland but the exploitation of coastal resources by occupants of temporary camps was a key feature for these communities at certain times of year.[35] This is confirmed by Melanie McQuade, who has excavated the remains of fish traps made of woven wattle fences containing mainly hazel rods, within what was then the Liffey estuary. These fences were set on the shoreline to catch fish swimming in and out with the tide.[36]

Today, Dalkey Island is a destination for visitors from local towns and from all around the world. People are taken across the short stretch of water to the island by the ferryman, Ken Cunningham. He is a great ambassador for the island, as he gives all the visitors an introduction to the history and wildlife of the area. Here, just a short train ride from the city centre, they find a natural coastal landscape, peace and quiet and beautiful views of the entire sweep of Dublin Bay and

Killiney Bay. The beauty of the latter bay, viewed from the summit of nearby Killiney Hill, is often likened to the Bay of Naples.

Shanganagh Cliffs

Viewed from the sea, the southern part of Killiney Bay is backed by a line of low cliffs composed of boulder clay, and these are rapidly being eroded by the waves. This was the scene over a century ago of a remarkable find that puzzled many naturalists of the time. Praeger walked this shoreline in 1896, and he takes up the story:

> One day in February last, Mr. R. Welch and I strolled along the beach northward of the new harbour at Bray, and just within the confines of the County of Dublin. At the verge of low water, where the slope of coarse shingle gives way to a more level stretch of fine sand and boulders, which is only left dry at spring tides, we noticed some stumps and boughs of trees, and on examining them, found that they were embedded in a compact layer of peat, which dipped southward at a low angle. The peat was

> full of branches and roots, and of cones of the Scotch Fir. On the southern side it disappeared under a bed of fine blue clay containing sea-shells; to the north, its broken edges overlay a stratum of coarse grey sand, with rounded fragments of granite. We had but cursorily examined the spot when the tide crept up again and soon hid it from view. Here evidently was a geological story to be unravelled; a long history lay buried with this old peat-bed under the mud and shingle which the sea had heaped upon it; and it was for us to read that history, if we could.[37]

What Praeger and Welch had stumbled upon was in fact a prehistoric forest of Scots pine with the timbers embedded in peat. The trees could either have grown on the site or logs may have been washed downstream and become embedded in sediments in a delta. Even the presence of peat on a submerged shoreline suggests settled terrestrial habitats at a lower sea level than present today. As would be expected of two of Ireland's finest naturalists of their day, they made a detailed description of the site which was published a few months later in the journal *The Irish Naturalist*.

Even 125 years ago, Praeger understood that this demonstrated the relative changes in land and sea levels since the last Ice Age. Much more recent technology using radiocarbon (C^{38}) dating has enabled a precise date of 8,600 years before present to be estimated for the submerged timbers. With a much lower sea level at that time I can picture the areas now covered by shallow seawater to have been covered with woodland when the first humans arrived in Ireland. I took a walk along this shoreline at the very lowest point of the tide to see if I could rediscover Praeger's 'pine forest'. But extensive erosion over the last century has already removed much of the beach, and there was no sign of this fossilised landscape.

The Murrough

Sailing past Bray Head and into Greystones Harbour I feel at home, as this is where I berth my yacht for most of the year. The harbour at Greystones was where my father's family went by train for their summer holidays in the 1920s, competing in swimming races in the harbour and learning to sail in small wooden boats. Now the quaint old harbour, with its rows of coloured fishing boats pulled up on the beach, has been replaced by a modern marina and a massive harbour wall. It is still popular today with crowds of day-trippers and local people using the small beach and the harbour walls throughout the year. It is a great place to learn to sail, with the great bulk of Bray Head and the Sugarloaf mountains providing shelter from the strongest of the westerly winds. I once noticed a cluster of seabirds off the harbour, with gannets and gulls diving into the water in a feeding frenzy. Then, without warning, a large minke whale surfaced among them, breaching several times as it joined the birds feeding on a shoal of fish.

Just to the south of the town I often walk along the top of the shingle beach at Kilcoole in summer where it is difficult to ignore the calls of the little terns that nest here among the stones. The smallest of the five species breeding in Ireland, they arrive here in May from their wintering grounds in west Africa. I can pick them out by their dainty, moth-like flight. Each pair lays up to four eggs on the ground but the nest is just a depression in the shingle and the birds rely on camouflage to protect their eggs and young chicks from predation. However, unintentional disturbance by walkers and dogs on the beach can be fatal if the adults are kept away from the nests and the eggs robbed by the ever-present crows or gulls.

Back in the 1980s, I was working for the Irish Wildbird Conservancy (now BirdWatch Ireland) and in the summer I often met local birdwatchers here. We were all concerned about the level of disturbance to the terns on the beach so a group of us decided to mount a voluntary watch on the site, to erect some signs and to approach regular walkers here asking for their cooperation to help the terns during the few short weeks that they spent on the beach. Several members of the newly formed Wicklow Branch of IWC spent many days here and established a wardening scheme that could benefit the terns.

By 1990, we had purchased a caravan and this was moved to the site during the summer so that the wardens, now paid for their work, could look after the colony for 24 hours a day. It was quickly discovered that foxes were raiding the nests at night and cleaning out many eggs and chicks. Electric fencing was erected on the inland side of the colony with plenty of signs to explain the purpose of these measures to visitors. As the survival of the chicks improved and the colony grew over the years, other scavengers and predators were attracted to the site. Hooded crows gathered to steal the eggs, kestrels developed a taste for the chicks and even a peregrine was seen hunting the flying terns.

Occasionally, the sea also took its toll on the colony. Whenever a high spring tide coincided with strong onshore winds there was a risk that nests, especially those further down the beach, would be swamped by seawater. Anticipating this, the wardens were sometimes able to move the eggs up the beach in advance of the high tide, in the hope that the adult terns would continue to incubate. Such extreme conservation measures were justified as the Kilcoole colony had become the most important single breeding site for little terns in Ireland.

By 2020, the conservation work on the little tern colony at Kilcoole had been in operation for over three decades, supported for most of that time by the National Parks and Wildlife Service. The breeding population of little terns was one of the highest levels on record, with over 150 pairs making many breeding attempts that season.[39] I regularly sail along this coast in summer and it is always a pleasure when the boat is surrounded by these dainty seabirds dipping and diving into the waves.

The beach at Kilcoole is part of a large coastal feature on the Wicklow coast known as the Murrough. This is a complex of shingle beaches, lagoons and wetlands that stretches for fourteen kilometres along the coastal strip from Kilcoole Station in the north to Wicklow town in the south.[39] When I first visited the Kilcoole Marshes on the coast of County Wicklow in the 1980s there was regular shooting here and bird numbers in winter were small. Now the area is fully protected and flocks of lapwing and wigeon arrive from northern lands while greylag and brent geese graze the fields. I often watch otters here as they play and hunt in the shallow brackish waters that fill the lagoons. In summer the shingle here is bright with coastal plants such as yellow horned-poppy, sea campion and bird's-foot trefoil. Oystercatchers breed here and ringed plovers lead walkers away from their eggs at the top of the beach with their classic 'broken wing' display.

This is one of the best birdwatching areas within easy reach of Dublin City. In the centre is BirdWatch Ireland's East Coast Nature Reserve extending to over 92 hectares. It includes wet grassland, fen and wet birch woodland. As well as being important in its own right, the reserve is becoming increasingly valuable for wintering waterbirds, largely due to the management programme that was implemented here. The reserve was opened to the public in 2009 following funding support from the EU LIFE Nature programme that enabled BirdWatch Ireland to purchase the reserve and carry out a range of habitat restoration work, including wetland creation and woodland management. A key part of the work involved installing controls on water levels in order to restore the wetland habitats that had been whittled away through agricultural drainage over the decades. Shallow pools were created and reedbeds reinstated to give cover and feeding to birds like little grebe and grey heron.

When I walked along recently beside the railway, flocks of whooper swans cruised in to land on the floodwaters with loud 'whooping' calls. Here they joined thousands of ducks including wigeon, teal and shoveler and waders like curlew and black-tailed godwits, all feeding intensively in the wet ground. Then, out of nowhere, a hen harrier appeared quartering above the reeds that fringe the wetlands. It was hunting for small songbirds and occasionally it dropped out of sight, presumably intending to strike a victim.

Since the reserve opened a network of paths has been created with wooden boardwalks over wet ground, and there are state-of-the art birdwatching hides for visitors. Interpretation boards along the trail explain what can be seen and why the reserve was created. As well as local people and individual birdwatchers there are many visits by schools and other groups. Numbers of visitors have increased, so there is a need for a good car park on the narrow, winding road to the main entrance. BirdWatch Ireland volunteers help to maintain the facilities. In its first ten years the reserve has been an unqualified success, restoring a coastal wetland and attracting a diverse community of waterbirds.

At the reserve, I met Eric Dempsey, local birdwatcher and writer. He said, 'the East Coast Nature Reserve has been changed from a dry area into a superb habitat with flooded fields, extensive reedbeds and woodlands. The diversity of bird species found throughout the year is growing as the reserve matures. To have such a fantastic reserve so close is special, not just for me as a birder, but for the community, schools and the many visitors enjoying the facilities that the reserve has to offer. For me this is a dream come true.'

Wicklow Head

Wicklow Harbour is the location for the start and finish of the biennial Round Ireland Yacht Race. Described as Ireland's premier offshore yacht race, it is the second-longest race in the Royal Ocean Racing Club calendar. Circling the entire island of Ireland in a clockwise direction, the yachts cover just over 700 nautical miles (1,100 kilometres) in a race that can include everything from gales to doldrums. Organised by Wicklow Sailing Club, the first race took place in 1980 with only sixteen boats. In 1982 a full gale lashed the seventeen-strong fleet and all but a few boats ran for shelter.[40] Since then, the fleet has grown steadily, and now attracts entrants from all over the world. The fastest time for the race, at just over a day and a half, was set in 2016 by the Sultanate of Oman's flagship trimaran.

Wicklow Harbour is where I used to moor my own boat but I moved it permanently to Greystones after a north-easterly storm tore it from the moorings and threatened to wreck it on the beach. I know these waters very well, having many times sailed out to the Codling Bank and the Arklow Bank offshore. The biggest challenge is the strong tidal stream that powers past Wicklow Head on both a rising and falling tide. The headland sticks out into the Irish Sea, making it the most easterly part of the Republic of Ireland, and the tide moving up the Irish Sea is forced to flow faster, just like a river flowing around an obstacle in its path.

This fast tidal race is one of the reasons why the seabed at this location has an unusual reef made entirely by the honeycomb worm *Sabellaria alveolata*. The animals live in small tubes that they construct by cementing together coarse sand and shell material. The tubes are arranged in big clusters and look something like the inside of a wild bee's nest where the insects have built their honeycomb in random shapes. These reefs range from thirty centimetres to two metres in height and take the form of hummocks, sheets or, sometimes, massive formations. On their own these structures are interesting but the diversity of other marine animals which live in nooks and crannies of the reef, somewhat like coral reefs in tropical seas, make them a hotspot for marine life. This includes hydroids, polychaete worms, molluscs, bryzoans, barnacles, amphipods, crabs, starfish, brittlestars and sea squirts. The Wicklow Head Reef is of high conservation value, as it is the only documented example in Ireland of shallow-water reef built entirely by marine animals.

The headland is also a mecca for grey seals, which breed mainly in a number of large caves carved out by the sea under the contorted mica schist rocks. My curiosity drove me to explore one of these dark, mysterious places to find out where the seals were giving birth. I lifted the outboard engine on my small inflatable boat, rowed over the shallow rocks at the entrance and into pitch blackness. The sea was flat calm that day, and the tide was still dropping, so there was no risk of getting trapped. Nevertheless, a light swell kept the sea surface moving and ripples slapped against the sides of the cave. I stowed my oars and, with a strong head torch to show the way, used my hands on the seaweed-covered rocks to push the boat deep into the recesses of the cave. I could hear the plaintive cries of at least one seal pup calling for its mother from the boulder beach at the back of the cave. Then, in the torchlight I saw it – a small, weak-looking pup with a pure white coat. The adult seal slipped silently into the water and, with the speed and power of a torpedo, flashed beneath my boat and into the safety of the open sea. She hung about at the mouth of the cave waiting anxiously for the chance to return to her pup, so I retreated and left them both in peace. The pups are born here in September and October at a time when storms can drag them off the open beaches, so they are relatively safe in the higher parts of these caves. In a few weeks, after the pups become independent, large haul-outs of up to a hundred animals gather on the storm beaches of Wicklow Head, where they lie moulting and regrowing their mottled fur.

Above one of the largest caves on Wicklow Head stands the modern lighthouse. The light was automated in the 1980s and some of the keepers' houses have been sold to private interests. A much older lighthouse still stands on the summit of the headland. Octagonal in plan, it originates from 1781, when it was built as part of a unique pair of twin lighthouses. At the time there were only two other lighthouses on the whole east coast, and it was considered that a single light might be confused by approaching seafarers with either of the lights on Hook Head to the south or Howth Head to the north. This was a route that ships would use approaching the Irish coast to pass safely between the shallows of the Arklow Bank to the south and the India Bank to the north. However, the upper light was often obscured in foggy weather, and so a third tower was built in 1818 much closer to sea level and the upper building was decommissioned.[41] Today, the older tower is a comfortable apartment managed by the Irish Landmark Trust and available for visitors to rent. On a clear day it gives fine views of the coastline to the north and south and of the Wicklow mountain range to the west.

Arklow

Sailing down the east coast past Wicklow Head with my friend Cormac, we passed by a line of seven large wind turbines, the only ones to be built so far on the seabed off the Irish coast. I have sailed up so close to these huge structures that the rotors seem to be turning slowly almost overhead. Commissioned in 2004, the development on the Arklow Bank was intended to be the first phase of a huge windfarm with 200 turbines but delays with onshore facilities and licencing have so far prevented further building. Each of the existing turbines reaches 124 metres from the seabed to the top of the blades or about the length of a standard Gaelic football pitch. I have to be careful, though, because the turbines were built on the crest of a shallow submarine ridge of sand and gravel. Praeger remarked:

> The great banks of sand and gravel that lie some miles out at intervals all the way from Dublin Bay to Carnsore Point represent the wreck of a land made of glacial drift. These banks have been moulded by the tides so that they form narrow north-south ridges, and wave action maintains their crests at a very uniform depth of one to two fathoms, while the surrounding water is deep enough to float the largest vessel. Lightships on their seaward side warn vessels off, for the banks are a mere death-trap. If a ship grounds, the tide digs a deep hole on the down-stream side, into which the vessel topples and in a surprisingly short time she disappears completely.(WW)

Suitably warned, I entered the port of Arklow between two narrow training walls that reach out into the Irish Sea carrying the sediment-filled waters of the Avoca river. From its tributaries, the Avonmore, Avonbeg and Glenealo, to the sea this river drains most of the east side of the Wicklow mountain range. Tying up to a pontoon alongside the river walls, I can see immediately that the water quality here is not good. This is still one of the few coastal towns in Ireland without a modern sewage treatment plant, although plans are well advanced for a new facility located in the redundant industrial area on the north quay. However, the project will take a number of years to construct.

Unfortunately, this river also has a long history of industrial pollution, largely emanating from the Avoca Mines which operated some ten kilometres upstream. There has been mining in this area since pre-Christian times, but the first modern record of a copper mine here dates from the end of the seventeenth century. By the 1840s sulphur became the main focus of the mine operations. The harbour at Arklow was still undeveloped at this stage, with ore being transferred by small vessels to larger ships, lying at anchor offshore, before export to Britain.[42] Mine waste was pumped into large settlement ponds, some just upstream of Arklow town, and the legacy of these is the toxic residue of heavy metal that continues to find its way to the river, causing damaging impacts on the marine and freshwater life. James Wilson, of Trinity College Dublin, studying marine molluscs here, found that some species were unable to reproduce at all.[43] For decades this river and its estuary were among the most polluted in Ireland, but things have been slowly improving with the rehabilitation of the mine waste.

Arklow has a long maritime history, and was once an important fishing town and boat-building centre. As I walked down some of the narrow lanes that lead from the main street to the river, I could imagine hundreds of small sailing boats lining the quays, a bustling community of fishermen and fishsellers, women working on the manufacture and repair of nets and timber boats under construction. Since at least the 1770s, the Tyrrell family designed, built or skippered boats out of Arklow. The shipbuilder John Tyrrell founded a family business here in 1864 and, after building dozens of fishing smacks and other sailboats, he and his son built the firm's first motor-powered yacht in 1904. I walked up to the Arklow Maritime Museum on the North Quay to learn about the building here in 1981 of the traditional sail training ship *Asgard II*, which eventually sank in the Bay of Biscay in 2008.

Leaving my boat in Arklow, I walked along a beach called Enerreilly north of the town where the Redcross River meanders into the sea. The beach here is made up of flat stones interbedded with large quantities of empty oyster shells. I used to collect these from various beaches in south Wicklow and, after smashing them with a sledgehammer, gave them to our poultry who needed calcium to make their eggshells stronger. I wondered why the empty shells were so numerous on these beaches as there are no native oysters alive today on the east coast. According to Noel Wilkins, a marine biologist in Galway, there were enormous and prolific oyster beds off the coasts of Wicklow and Wexford up to the early 19th century, and these are shown on a number of ancient maps of the Irish Sea.

The increasing urban populations of the early 1800s led to a massive increase in exploitation of oysters close to the city of Dublin and, for a while, these were restocked with shellfish from the Arklow beds. Traditionally these had been dredged by oystermen from Arklow and shipped to Liverpool for the English markets. In the 1840s about a million oysters were exported each year from this port alone to restock the depleted beds on the English Channel coast which had been supplying the important London markets. Twenty years later the landings in Arklow had reached unsustainable levels of 30 million oysters each year, and it was not long before the local beds became depleted themselves. By the end of that century the beds had been commercially exhausted and never recovered.[44] This classic case of overfishing remains a sad tale from the Irish coast and accounts for the large quantity of empty shells on the beach, most of which must have been in the sea for more than a century.

Wexford Harbour

Passing the long, straight coastline of north Wexford, I can see the low cliffs near Morriscastle, where the sea has eaten away at the soft sediments and where there is a classic example of a building that has partially collapsed into the sea. The afforested peninsula of Raven Point marks the entrance to Wexford Harbour, which is one of the largest estuarine areas on the east coast. Sailing into the sandy harbour is hazardous due to shallow water, but fishing trawlers and small cargo vessels do moor alongside the quay walls in Wexford town. The wider bay is a confusing maze of shallow sandy banks and channels which was once twice as large as it is at present.

The Wexford Slobs were reclaimed from the sea in the 1840s and closely resemble some of the polders found in the Netherlands. Two long sea-dykes were built enclosing a total of over 2,000 hectares of mud and sandflats which were then converted into grassland with some cereal and root crops. To maintain this farmland in a dry condition, surplus water must be continually pumped out to the harbour. In 1969 the Wexford Wildfowl Reserve was established on the North Slob and today this holds a large portion of the world population of Greenland white-fronted geese and pale-bellied brent geese, together with a wide range of other waterbirds. This is also one of the best places in Ireland to see wild Irish hares, which seem to prosper on the large open fields that are mostly lacking hedges. Just north of Wexford town, I went to the viewing tower at the visitor centre, which gives ideal views of the birds on the nearby ponds and fields. I sat on the sea wall as the sun set over Wexford town and watched long skeins of geese fly out from the North Slob to their night-time roosts on the sandbanks in the harbour.

The shallow waters of Wexford Harbour are an important place for the bottom culture of mussels. The shellfish are generally grown from juvenile mussels (or seed) dredged elsewhere and laid out on the sandbanks in the harbour. After several years the full-grown mussels are harvested with up to 7,000 tonnes landed and sold here each year.[45] The spat (or juvenile shellfish) are released by mature mussels into seawater in millions each year, and these settle on suitable stretches of seabed. Shortage of spat has been a problem in recent years and, despite an increase in seed mussel fishing in the Irish Sea by a fleet of licensed dredgers fitted with underwater cameras, the present annual catch of around 20,000 tonnes only provides about a third of the actual seed needed. I have often seen the slow-moving dredgers working along the Wicklow sandbanks with the teeth of the dredges scraping the seabed and collecting everything in their path, including much bycatch. In the last decade surveys by Bord Iascaigh Mhara, Ireland's seafood development agency, have recovered very few seed mussels here, and in some cases report only large numbers of starfish and dogwhelks, both predators of mussels.[46] This looks suspiciously like another case of overfishing, with the seabed no longer suitable for spat settlement.

Guarding the south side of the entrance to Wexford Harbour is the long, low sandy spit of Rosslare Point, which has been broken up several times in the past by easterly storms. In the 1950s, sand from the point redistributed inside the harbour and gradually built up into a dune which became known as Tern Island.

The island was unstable and continually changed shape but despite this it quickly became the largest tern colony in Ireland. As I related in the introduction, I went here several times in the 1970s to assist with a research project, and the cacophony of sound from thousands of seabirds was impressive, with all five Irish breeding tern species and a large number of black-headed gulls nesting here.[47] Several years later the island was completely washed away in storms, and the birds redistributed to Lady's Island Lake on the south Wexford coast and to Rockabill island off north Dublin.

Carnsore Point

As I sailed from the north towards the south-east corner of Ireland the most prominent landmark was the distinctive outline of Tuskar Rock. The massive granite lighthouse stands high on a flat-topped rock to guide ships past the dangerous rocks off Carnsore Point. However, this place has a history of tragedy. Building of the lighthouse began in 1812, but on a night in October that year a sudden, severe storm blew up, whipping the sea into a cauldron of enormous waves. The rock is located exactly where Atlantic swells meet tides coming down the Irish Sea and gales from the Bay of Biscay. In the days before accurate weather forecasting there was no warning, and waves swept over the rock, tearing away the wooden dormitories of the twenty-four workmen. Ten men were drowned, although the remainder survived by clinging onto the rocks for two days until they were rescued. Incredibly, all the survivors returned to continue their construction work and the lighthouse began its service three years later.

Since then the lighthouse has guided ships into the Irish Sea for over two centuries, but on just one occasion it failed to be lit up. This coast was a favourite location for smugglers to land contraband shipped in from French ports. In 1821, the cutter *Shark* was bound for Wexford town under cover of darkness with a cargo of French brandy, candles made from sperm oil and several other commodities that were taxable. When the crew saw a coastguard cutter patrolling the entrance to Wexford Harbour, they landed the boat in a cove on Tuskar Rock and entrusted their illegal load to the lighthouse keepers on the promise of a cut of the goods on their return. When the *Shark* approached the Tuskar a few nights later, the crew were surprised to find the lighthouse unlit and the two lighthouse keepers totally drunk and asleep. By complete coincidence, on that very night the *Royal Yacht* carrying King George IV was passing the area accompanied by a Royal Navy escort bound for Dublin Bay. Failing to find the Tuskar light, the yacht ran for shelter on the Welsh coast instead.[48] Despite the presence of the lighthouse, an American ship was wrecked on the nearby Blackwater Bank in 1859, with 424 passengers and crew being drowned. When divers later examined the sunken wooden hull they found nearly 300 corpses below deck, suggesting that they had been trapped here as the ship sank.[49]

In 1968 this was also the location of a major air accident that puzzled investigators for many years afterwards. Aer Lingus flight EI712 was on route from Cork to London when it disappeared into the sea just off the coast of Wexford near Tuskar Rock. All fifty-seven passengers and crew died, making it the single biggest loss in the history of Irish aviation. Rumour surrounded the accident for many years. The most popular theory was that the aircraft had been fired on during a British military exercise in the Irish Sea, and that this had been covered up by the London government.

More than thirty years later, detailed forensic work by an expert international team concluded that this was simply an accident caused by damage to a tailplane shortly after takeoff, which sent the aircraft into a fatal downward spiral.[50]

There is a traditional boat associated with the Wexford area called the 'Rosslare cot' which, despite its small size, was a seagoing craft. Its clinker-built planking and pointed bow and stern may even have evolved from the boat-building methods of the Vikings who settled in Wexford in the 9th century. These boats, which were between nine and twelve metres long, ranged over the Irish Sea. During the 1920s as many as forty-six large cots were fishing out of Rosslare, each crewed by four men. In settled weather some Rosslare fishermen would row out the sixteen kilometres of open sea to Tuskar Rock, where line fishing was profitable.

The other main landmark in this area is the large windfarm on Carnsore Point, just south of Rosslare. This was the site of a controversial proposal to locate Ireland's nuclear power station here in the 1970s. The plan was for one, and eventually four, nuclear power stations, but anti-nuclear groups organised a series of rallies and concerts at Carnsore Point in protest.

The gatherings were accompanied by a big publicity campaign that brought the whole issue of nuclear power in Ireland into the public eye. The proposal was dropped in the late 1970s and the ESB eventually used the land for one of Ireland's first renewable energy projects.

These fourteen wind turbines are still among the few that are located close to a beach in Ireland. I know this place intimately, as I spent two years carrying out research here to discover if there were any negative impacts of the windfarm on the shorebirds using the nearby beach. Migratory godwits, curlews, oystercatchers, dunlins and a number of other wader species feed each day within the shadow of giant turbine blades rotating in the almost constant wind. Using a strong telescope, I was able to count the number of worms or shellfish eaten per minute by a large sample of birds and compare this with rates measured on similar beaches where there were no turbines. My clear finding was that the behaviour of the birds was unaffected by the turbine activity, and that their feeding rate, and hence their chances of surviving the winter and the long return migration to the arctic, was dependent mainly on the density of their prey beneath the sandy beach. With great political pressure now to increase the contribution of renewable energy to the national grid, there will be a need to locate more windfarms on the coast in future, but great care needs to be taken in siting them where they will cause least disturbance to wildlife.

Leaving the East Coast

Despite being overshadowed by the more famous Wild Atlantic Way, the east coast of Ireland has many places of interest, and some which I have yet to discover. This coastline is relatively straight and low-lying, with fewer islands than the rest of the country, although there are many estuaries that are important links in a chain of migratory bird habitats. More sheltered from the Atlantic winds and wave energy which come mainly from the west, it provides ideal conditions for major ports such as Belfast, Dublin and Rosslare, with the largest yachting marinas in Ireland at Bangor, Howth and Dún Laoghaire. With the cities of Belfast and Dublin situated on sheltered bays, it is more urban than the other coastlines, and has a history of extensive trade with Britain and Europe. There are strong family links here too between Ireland and the facing coasts of Scotland, England and Wales. The Vikings first landed on the east coast in the 9th century, and they left behind a legacy in the placenames such as Strangford, Carlingford, Wicklow and Wexford. The east coast offers unlimited space for casual, beach-based recreation and more formal water sports such as sailing.

Praeger was born near Belfast and lived for most of his life in Dublin, where he explored and documented key elements of the natural history of the east coast. Living within sight of the Irish Sea for most of my life too has given me an intimate knowledge of its inshore waters and varied shorelines. Taking care to navigate safely through some dangerous rocks off Carnsore Point, I leave the east coast and set sail along the start of the south coast with the Saltee Islands in sight.

In the 1880s Praeger became frustrated with the inconsistent nature of his work as a freelance engineer on various public schemes such as drainage projects and harbour development around Belfast. Meanwhile, he had become an accomplished naturalist and active member of the Belfast Naturalists' Field Club. It was obvious by this stage that his life's work would in some way be dedicated to the study of Ireland's natural environment. In 1888 he was unsuccessful in his application for a job in the National (then the Science and Art) Museum in Dublin. His desire to change direction from engineering to natural history was noticed by the curator of the museum, Robert Francis Scarff, who suggested that he apply for a position as an assistant librarian at the National Library. Although he had no qualifications or experience in this profession, he accepted Scharff's advice that 'librarianship was largely a matter of general knowledge and common sense'.[1]

Despite leaving the area of his youth, Praeger relished the move to Dublin, continuing his involvement with naturalists' field clubs. He explained, 'When I left the north in 1893, I exchanged the Belfast Club and Ulster for the Dublin Club and Leinster, Connaught and Munster.'(WW)

Working in Dublin at the National Library, Praeger was surrounded by the scientific institutions that allowed him to mix with all the key naturalists of his day. These included the Royal Dublin Society and the National Museum, both in the Kildare Street 'just across the grass plot dominated by the uninspiring statue of Queen Victoria,'(PS) the Royal College of Science next door and the Royal Irish Academy in nearby Dawson Street. In fact, while resident in Belfast, he had already been elected a member of the Academy in 1892. He wrote, 'for a young man interested in science, as I was, the situation was ideal'. His excitement was palpable. 'It was an exhilaration and a delight to me to have my limited stock increased daily by fresh facts and fancies bearing on so many branches of knowledge.'(PS)

Praeger's exploration of the provinces of Ireland was assisted by the fact that the network of small railways at that time was much more extensive than it is today, reaching to remote places such as Dingle, County Kerry; Lahinch, County Clare; Clifden, County Galway; and Letterkenny, County Donegal. He was able to catch a train to almost any part of Ireland to do his fieldwork. At that time the rules of the Civil Service were applied with a 'wise liberality', especially in the National Library. It was possible for Praeger to spend long weekends in the country, from lunch on a Friday till lunch on the following Monday. In this way he managed to travel for a total of 200 days in the field over the last five years of the nineteenth century, walking for more than 7,000 kilometres and recording wild plants in every county in Ireland for his monumental work, *Irish Topographical Botany*.

Praeger was modest in his accommodation needs and, if far from town or village, he would happily sleep in the open on a bed of soft heather or bracken. He shunned the large tourist resorts, which he considered should be, 'evaded by humble-minded naturalists like myself, as suiting neither our pockets nor our temperament. Personally, I like a homely small hotel, where there are no gongs or waiters or pygmy coffee-cups; where your host is full of information about local matters, and the son of the house wants to sail you out to foamy skerries to see the seals and the seabirds'. (WW)

There are plenty of seals and seabirds on the south-east coast, which Praeger described thus: 'The Irish coastline here at its flattest .... in Wexford great stretches of sand and gravel, backed sometimes by shallow inlets of the sea'.(WW) In fact, the beaches here are so extensive that they are often deserted even in the midst of summer. Streams of seabirds flew past me as I sailed around Carnsore Point. Ahead of me I could see the long, straight coast and the familiar outline of the Saltee Islands.

Lady's Island Lake

One of the shallow inlets referred to by Praeger is Lady's Island Lake, named after a Christian pilgrimage site on an island in the lake. As sea levels rose further following the end of the Ice Age, the mouth of the inlet was filled with a barrier of sand and gravel blocking off the tide. It is now known that the pressure from wave action does allow seawater to penetrate through this barrier, and the salt concentration in the lagoon changes from south to north. Places where freshwater and seawater mix are usually described as brackish, while natural lagoons such as this are relatively rare on Irish (indeed, European) coasts, as a consequence of which they have been given special legal protection under the EU Habitats Directive. A major national survey in the 1990s found over a hundred such habitats in the Republic, ranging from natural saline lakes to artificial lagoons cut off from the sea by roads or causeways. The unusual salinity conditions and frequent fluctuations in levels in these lagoons support a unique collection of plants and animals, but most of this is below the waterline, and therefore normally out of sight.[2]

At Lady's Island Lake, the sand on the barrier itself is home to one of Ireland's rarest flowering plants, cottonweed, which survives here precisely because of the instability of its habitat. I have been present here just after a large, tracked excavator opened a channel through the barrier, allowing lagoon water to pour out at low tide. This annual cutting is normally done in the late winter or spring, when water levels are highest. Standing on the edge of 'the cut' is a dangerous activity, as the sand can crumble at any time from the power of the water gushing out to sea. After a number of weeks, the gap is normally closed by the action of the sea on the shingle beach and the inlet becomes a lagoon again. This management is undertaken by the NPWS so that around April the islands in the lake are available but surrounded by water just in time for the return of the terns. These small migratory seabirds will have spent the winter in African waters and are here to breed. The islands in Lady's Island Lake hold the largest colony of Sandwich terns in Ireland, at around 1,700 nesting pairs. The birds require extensive, sheltered, shallow waters with a supply of small fish to support young chicks and adults.[3]

I first visited these islands in the company of Clare Lloyd and the late Oscar Merne, both experts on all seabirds, as part of a study of migration and survival of the terns. Recent studies of the terns have shown that they are back in west African waters by October, principally on the coasts of Senegal, The Gambia and even as far south as Namibia. Their annual switch from one hemisphere to the other means that they live in an almost continual summer, something we would all like to do at some time.

Saltee Islands

West of Lady's Island Lake there is a long, unbroken coastline of sand and shingle which also encloses the lagoon of Tacumshin. Arriving by sea at the small fishing harbour of Kilmore Quay is usually a pleasure, because the marina in one corner is welcoming and well run. On one occasion, the engine on my yacht broke down as I approached with my friend Karl Partridge and, because the channel outside the harbour is narrow and bordered by rocks, the local lifeboat was dispatched to tow us safely into a berth. From the harbour, Declan Bates runs a ferry service to the nearby Saltee Islands in his two boats, named after the razorbill and guillemot. Located close to major trading routes on the south and east coasts of Ireland, the islands became a base in previous centuries for pirates and smugglers who regularly plundered ships that were wrecked or driven ashore in storms. A common practice was to tie a lighted lantern to the horns of a grazing cow so that the moving light looked like a ship at sea, thus luring cargo ships onto rocks or beaches, where they were robbed. The caves on the south side of the island were an ideal location for hiding plundered goods.[4]

These islands have long held a fascination for

ornithologists due to their large seabird colonies. Praeger described the larger of the two islands, Great Saltee, as 'one of the most populous and interesting breeding grounds of sea-birds to be found in Ireland'. He recalled:

> I made the acquaintance of the wonderful avifauna of the Saltees during a pleasant week in 1913, as one of a party of zoologists and botanists. The birdmen were desirous of studying the night-life of this populous city, so we disclaimed the clock and all its works, and came and went as suited us, by day or by night, sleeping on a wisp of hay on the floor of the farmhouse (and getting plenty of fresh air since half of the roof was gone) or out among the bracken on the hill-side.(WW)

I have slept in the same farmhouse, now restored and comfortable. As I recount in the introduction, I first went there in the 1970s as one of a party of biologists to study the seabirds, and I have been on many summer visits since. Apart from the condition of the farmhouse, the comparisons with Praeger's time are stark. The island was still grazed during Praeger's visit, and had been farmed since at least the 16th century until the family here took to smuggling as a more profitable business. The botanist Henry Hart visited in the 1880s and reported that 'on this island which is partly cultivated there is a resident family. The Lesser Saltee is used as pasturage and contains but one cabin for the use of herd boys.' There are still the remains here of stone-built circular platforms that were used for drying corn prior to threshing. The farmhouse was lived in by a family up to the 1940s, when they were forced to abandon farming. Great Saltee was later sold to the self-styled Prince Michael Neale, and his descendants still own this private sanctuary. In 1950 the Saltee Bird Observatory was founded here by Robert Ruttledge and John Weaving. Lying off the south-east coast of Ireland, it was an important stopping point for exhausted migratory birds arriving from the European continent in spring and awaiting favourable winds in the autumn.[5]

Praeger noted in 1913 that an interesting new seabird on the island was the fulmar, which had been first recorded breeding in Ireland only a few years earlier. By 2018, this oceanic species, which is closely related to the albatross family, was breeding all around the Irish coast with a population estimated at 33,000 pairs. Similarly, Praeger noted in 1913 that the gannet was another newcomer, and that 'for some years now one or two pairs have bred on the Great Saltee'. By now the population of gannets here has expanded to reach almost 5,000 breeding pairs.[6] On my recent visits, noise from their colonies was deafening and, as I sat at the top of steep cliffs on the south side of the island, the smell of their guano was equally impressive. The gannets are large and very noisy, with a constant patrolling of adults around the colony – a feast for all the senses. The Saltee Islands are also an important breeding site for grey seals. This assembly is known to exchange individual seals with other colonies in the Irish Sea, including some on the Welsh coastline. In the late summer, when moulting is underway, many hundreds of these animals haul out in the nearby Wexford Harbour.[7]

Hook Head

Sailing past the long, narrow promontory of Hook Head, the most obvious landmark is the squat lighthouse with its distinctive black-and-white rings. This is thought to be the oldest working navigational lighthouse in Ireland or Britain, dating originally from the 5th century, when it was little more than a fire stoked by monks in order to warn passing sailing vessels of the dangers of the rocks. When the Normans arrived in Wexford they built a fortress and watchtower in this strategic location. The present tower was built in 1172, and after 850 years it is still in excellent condition.[8] When I visited the lighthouse, I went inside and climbed the stone steps to the top from where there is a wonderful view of the coasts of both Wexford and Waterford. The tower is four stories high and has walls that are up to four metres thick. Richard Taylor, who worked on servicing the lighthouse, remembers one particular keeper called Bill Hamilton. 'Bill would have to pass my room on his way to wind the light – a mechanism that operated not unlike a grandfather clock, and which had to be seen to every 40 minutes. Bill's yarns were so long that by the time he had finished one, it was time to wind up again.'[9] Today Hook Lighthouse is open to the public and guides are on hand to explain the intricacies of running a lighthouse in previous centuries.

The bedrock at Hook Head consists of two types of sedimentary rock: old red sandstone and limestone.

A band of sandstone runs across the peninsula from Broomhill to Carnivan. For centuries this was quarried at Herrylock to make millstones, water troughs and other useful objects. The limestone rock was burned in the many limekilns, some of which can still be seen on the peninsula. The lime powder which this produced was used to improve the quality of the soil. It was also mixed with sand to make lime mortar for building stone walls and houses. The point of Hook itself consists of beds of limestone with some of the best-preserved fossils that I have ever seen in Ireland. These include the casts of corals, crinoids (sea lilies), brachiopods (shellfish), bryozoans and starfish, remnants of marine life in a tropical sea some 350 million years ago.

The rocky shore and seabed here are popular dive sites for scuba divers. There are exposed intertidal reefs and subtidal reefs, typically strewn with boulders, cobbles and patches of sand and gravel. The reefs around Hook Head have species-rich, tide-swept marine communities in both the shallow and deep-water areas. The rocks below low water mark are covered with dense kelp beds. In the deeper waters there are many colourful encrusting animals, such as cushion sponges, with branching sponges and rose 'corals'. Sea squirts and brittlestars add a bit of action to the seafloor. In gravel beds there are burrowing sea cucumbers. Rare seaweeds also add to the importance of Hook Head, which is now a protected European site.

A strong tidal flow passes the point of Hook Head, where it meets the outcoming tide from the large Waterford Estuary. Sailing up to Waterford City I was careful to catch the incoming tide, which carried my yacht along at a steady six knots. The wide waters are bordered by farmland sweeping down to small sandy coves or low rocky cliffs. On the Wexford side at Duncannon, a prominent headland composed of hard volcanic rock is the site of an impressive star-shaped fort that was built in 1568 in anticipation of an attack by the Spanish Armada. But the invasion of 1588 did not reach this area, and instead up to twenty-four Spanish ships were wrecked in storms off the west coast.

Dunmore East

Praeger must have felt that the coast of Waterford was underappreciated, as he devoted eight pages to it in his book *The Way that I Went*.

> I have lingered over the Waterford coast because it is known to very few and is well worthy of being known better .... Flowery slopes, lofty cliffs and stacks and pinnacled islets with colonies of seabirds of many kinds as well as choughs, peregrines and ravens; and an illimitable glittering sea extending to the southward. You can follow the shore tolerably closely by road but walking along the breezy clifftop is the way to enjoy it properly. There are not many places to stay ... but that is one of its charms, for it keeps it unchanged and unspoiled.

Today, there are plenty of places to stay in the holiday resort of Dunmore East, which lies on the opposite side of Waterford Harbour to Hook Head. There is no marina here as this is a busy fishing port, but some pontoons have recently been added to the north-west of the old lighthouse, and I tied up here for a short time with the harbourmaster's permission. As I walked around the harbour, between the usual heaps of fishing nets and old boats, my attention was drawn to the cries of kittiwakes on nearby cliffs. These small seabirds breed under the park at Outer Harbour and off the sea wall at Black Knob due south of the lighthouse. They previously nested in the Inner Harbour and I once had the opportunity to investigate these cliffs closely, using a mechanical 'cherry-picker', as there was a problem with soil slippage and I was requested to map the cliff vegetation. The kittiwake nests are amazing constructions containing seaweed and grass, cemented to almost sheer rock with the bird's own droppings. Many of the nests in the harbour colony also incorporate fragments of discarded fishing net. Because it does not rot, this synthetic netting material has accumulated layer upon layer over many years, and probably results in more stable nest platforms than would be usual in a more remote colony.[10]

While most of Ireland's seabird populations are in a reasonably healthy state, kittiwake numbers have declined by over one-third since the 1980s, and breeding success has fallen markedly.[11] Poor breeding performance has been linked to declines in food availability, which has also been explained by rising sea temperatures as a consequence of climate change. This may also be partly due to the reliance of the birds on small shoaling fish such as sand eels, sprat and young herring. The practice of pair-trawling of spawning inshore sprat has increased in recent years and, with a thriving herring fishery in the Celtic Sea, there are likely to be implications for the breeding success of Kittiwakes along these coasts.[12] This is ironic, given the close proximity of the Dunmore East colony to the fishery harbour.

Tramore Bay

The seaside town of Tramore lies on a fine stretch of sand that is popular for swimming and surfing. Praeger knew the 'line of steep dunes backed by salt-marsh running out from the eastern side and nearly cutting the bay in two. This is the home of some rare seaside plants such as Asparagus'.(WW) This wild ancestor of the popular vegetable is still present there today, although parts of the Backstrand at Tramore are now surrounded by artificial sea walls to prevent flooding of the town and neighbouring farmland. It is also the location of an old landfill dump that, from the 1970s on, threatened the conservation value of the shoreline for birds and plants. This dump has since been remediated and is now used as an amenity walkway. Remarkably, around 500 bee orchids were seen there recently, showing that nature can partially recover even in severely damaged habitats.

As compensation for the environmental damage caused by siting the town dump on the Backstrand, Waterford County Council was required by the European Court of Justice to create a new wetland habitat. Compensating for the loss of beaches and mudflats is a difficult challenge, but it is increasingly being addressed in other countries by a technique known as managed realignment. In many estuaries of Ireland tidal land was reclaimed from the sea in the 19th century by building sea walls and pumping the water out. In 2013 the Tramore Wetland Restoration Project began when part of the existing seawall was breached and several fields at Kilmacleague were inundated with seawater to produce 7.5 hectares of wetland habitat comprising mudflat, transitional salt marsh, upper saltmarsh and pioneer marsh.[13]

As I approached the wetlands with Bernadette Guest, Heritage Officer with Waterford City and County Council, a large flock of curlews rose from the margins, calling loudly, while some lapwings stood peacefully on the new mudflats. Aside from compensatory habitat creation, the Tramore wetland restoration project is the first example of managed realignment in Ireland. This is a management tool for estuarine habitat, whereby sections of flood defences are moved inland to create or restore intertidal habitat. It is increasingly being used in other countries as a 'soft-engineering' approach to combat the problem of sea level rise, storm surges and coastal flooding. The Tramore wetland restoration project therefore provides a good opportunity to study the success of such a project in Ireland. The monitoring of this site has been led by Lesley Lewis of BirdWatch Ireland. Within a three-year period, the created intertidal habitat at Kilmacleague had developed an invertebrate community, albeit of low species diversity.

Waterbirds have also taken readily to the created habitat, with a total of twenty-one species recorded in the winter so far, including brent geese, wigeon, little egret, dunlin, curlew, redshank and greenshank. They feed regularly within the created intertidal habitat. In the future, we can expect to see also oystercatcher, knot, black-tailed godwit, bar-tailed godwit that specialise on shellfish and polychaete worms. Given the quiet and undisturbed location of the Kilmacleague wetlands, it is surprising that roosting behaviour is not more important during rising and high tidal stages.

Bernadette feels that this project will act as a demonstration of what is possible in other areas to adapt to sea level rise. She expresses the personal satisfaction she has found with this project: 'In the 21st century where biodiversity loss is accelerating at a rapid pace it is a small win on a local level to see new habitat being created. It is rewarding to see the establishment of new saltmarsh habitat so quickly on the site and its use by large numbers of the wintering birds.'

Copper Coast

'West of Tramore there is twenty miles of almost unbroken cliff and precipitous slope tenanted by sea-birds and a wealth of wild-flowers, with small villages nestling in gaps of the rocky wall, where streams enter the sea,' wrote Praeger in 1937.(WW) From Tramore, I walked a section of this coastline along the line where the intensively farmed land meets rough grassland at the top of the cliffs. The calls of gulls were a constant sound along the cliffs, but occasionally I heard the distinctive cry of the chough, the red-legged crow of the Irish coast, in one of its most easterly breeding areas. Choughs use both types of habitat, the improved agricultural grassland in winter and maritime grassland with patches of gorse in the summer. This gives them the variety of feeding that they need to survive here throughout the year.[14] The presence of livestock, such as cattle, is a key factor, as it maintains a short grass sward for the choughs to probe. I also saw choughs flipping over cattle dung to find insects and other invertebrates, which are a vital part of their diet.[15]

The coastline of Waterford west of Tramore is known today as the Copper Coast. It is so named because of the mines that were an important part of the local economy during the 19th century. As I dropped down into some of the cliff-bound coves I could see in the rocks some of the metal-bearing veins that were mined all those years ago. The oldest rocks here were formed 460 million years ago during the Ordovician period. At this time, the rocks were part of a continental margin, located under the sea near the South Pole. As one continental plate slipped under another, magma rose from the depths of the earth's mantle until it erupted in two separate volcanoes. The first volcano erupted through the ocean floor. This created the dark-coloured volcanic rocks found along the coast and the mineralisation that produced the valuable metals.

The cliffs west of Bunmahon were mined for lead, silver and copper as early as the 18th century. However, the main phase of activity was in the mid-19th century, when the mines east of Bunmahon were worked by the Mining Company of Ireland. By 1840 this was described as 'the most important mining district in the Empire'. However, this proved to be the peak and, as miners were working at depths of up to 400 metres and a similar distance under the seabed, there was a serious threat of flooding. In 1850 the company began to move its entire operation, including engines and buildings, east to Tankardstown, where a new lode had been discovered. Hopes for an upturn in the price of copper and a new discovery in the area faded after twenty years, and the last few tonnes were sold from Tankardstown in 1879. The engines were sold for scrap and all that remains today of a mining operation that once employed 1,200 people are the ruins of the engine houses on the cliffs at Tankardstown. Entire extended families moved away, mainly to America, where some of them worked at the Copper Mountain in Butte, Montana. The mine buildings have been remediated to some extent and made safe, and interpretative panels have been erected at the site. There is a visitor centre in Bunmahon which outlines the mine's history.

Dungarvan Harbour

Dungarvan Harbour is only accessible to yachts at high tide, as the only available water at low tide is confined to a few narrow channels through the mud. I have sailed in a small boat here a few years ago and nearly ran aground on several occasions. At low tide, the yachts in the narrow town harbour sit on the mud, leaning at precarious angles. The bay outside is divided in half by a remarkable feature known as the Cunnigar, a long, narrow shingle ridge that has only one path on the crest. Praeger called it 'a straight narrow ridge of sand very curiously placed'. I walked out along the path, which has waves breaking on the eastern side and sheltered mudflats to the west. On the northern end of the ridge, a tightly packed flock of roosting shorebirds sat out the few hours during high tide until they could return to feeding on the mudflats. They included brent geese, oystercatchers, curlew, godwits and many other smaller waders. Their calls echoed across the bay to Dungarvan.

The outer bay, known as the Whitehouse Bank, is used by shellfish farmers and, at low tide, I could walk between the wire cages, known as trestles, that are packed with Pacific oysters. These are grown here until they can be harvested and exported to countries like France, Spain, Italy and the UK. Many Irish people seem reluctant to eat fish or shellfish such as oysters. For seafood in general, Ireland has been ranked just 48th within a group of 160 countries in terms of fish consumption per person. Perhaps it is a fear of the unknown, or maybe there is folk memory of the 19th-century famines, which forced starving people to forage on the shoreline. The oyster trestles are usually located in the lower part of the intertidal zone, in areas where the cages are covered with seawater most of the time. At Dungarvan, large blocks of trestles are spread across the sandflats on the outer beach. Oyster spat (or seed) is supplied by hatcheries and put in plastic mesh bags, which are placed on top of the wire trestles. Oyster husbandry activities mainly take place during low tides, when tractors drive out onto the hard sand. Their permanent tyre tracks are quite visible on the beach.

To many shorebirds, shellfish are a key part of their diet but, despite their name, oystercatchers do not feed on oysters but mainly on smaller shells such as cockles, which they break open with their stout orange bills, using the firmer sand as an anvil. Dungarvan Harbour was one of the locations for a landmark study by Tom Gittings and Paul O'Donoghue, who assessed the impact of the oyster-farming on waders. The birds showing a neutral or positive response to the trestles were waders such as turnstone that feed in small flocks or oystercatchers, curlews, greenshanks and redshanks that are normally widely dispersed or in loose flocks. They were able to move between, and in some cases on top of, the trestles. In contrast, knots, sanderlings, dunlins, godwits and ringed plovers tended to avoid the oyster trestles, as these birds mainly feed in large flocks of tightly packed individuals. So there were positive, neutral and negative impacts of the aquaculture on birds, depending on the feeding behaviour of each species.[16]

East Cork

The historic town of Youghal, situated at the mouth of the River Blackwater, marks the county boundary between Waterford and Cork. Praeger reminded us that the unusual English name of the town is an attempt to translate the original placename, *Eóchaill,* which means 'yew wood' in Irish. Although now largely silted up, Youghal Harbour was once an important port that traded goods with Bristol and many cities across Europe. Fish, timber and wool were exported, while glass, ironmongery, exotic spices and foodstuffs, clothes, wine and salt were imported. However, the arrival of the Black Death in 1348, and general political unrest, had a terrible effect on Youghal. But trade increased again and in 1462 this became one of the Cinque Ports of Ireland based on a charter given to the town by King Edward IV.

Youghal became the major centre for the export of wool from Ireland in the 17th century. Starting in the 18th century, when a great period of Georgian expansion began, the size of the town was enlarged to almost twice that of the medieval settlement. The merchants who leased parts of the waterfront were required to build new quays, outside the line of the old harbour and town wall. Other developments also occurred at this time, including the Clock Gate tower (completed in 1777) which is part of the medieval Town Wall. The quays that remain today, such as Green's Quay, Harvey's Dock and Nealon's Quay, commemorate the many merchants who built them.

By 1750 the old medieval harbour had been in-filled, and is buried today beneath the Market Square. The importance of the new port continued to grow, and by the early 19th century large quantities of timber, coal and anthracite were landed here, and substantial amounts of agricultural produce loaded for export. There was also a seagoing fishing fleet of 250 vessels.[17] With the large amount of shipping coming and going from Youghal, it was no wonder that many ran aground, or worse, on this largely unlit coastline. The Fleming family of Youghal were general merchants, with several ships involved in the coal trade between Bristol Channel ports and the home port of Youghal. In December 1913, one of these vessels, a three-masted schooner called the *Nellie Fleming*, was carrying a cargo of about 250 tonnes of coal bound for Youghal when she ran aground on the Black Rock off Curragh in Ardmore Bay. Some local fishermen rowed out to help while lifeboats from Youghal and Helvick also came to the vessel's assistance. The crew were saved but an attempt to refloat the ship was abandoned when water flooded her right up to the deck through a hole in the hull. A group of locals later set up a company to buy the cargo of coal in the wreck and resell it at a profit.

The valuable fuel was ferried by donkey and cart up to a field, where it was bagged, weighed and sold to local people. The *Nellie Fleming* now lies buried in the sand at the Ardmore end of Curragh Strand.[18]

The need for lighthouses to warn seafarers of hazards on this coast was recognised in the 1820s, and construction of a new one was begun in 1847 on the prominent Capel Island just off Knockadoon Head. However, when an important steamship, the *Sirius*, ran aground near Ballycotton in the same year, the location for the new lighthouse was changed to Ballycotton Head, and the stump of the tower on Capel Island was abandoned.[19] The new tower was completed on a smaller island at the entrance to Ballycotton Harbour in 1851. Here Praeger mentioned 'the little village of Ballycottin with its hotels tucked in below the hill out of the westerly winds, and its tall rockstack surmounted by a lighthouse'.(WW) There are many stories of heroic actions of local men in search and rescue, particularly the rescue of the crew of the Daunt Rock lightship which broke its moorings in a gale in 1936. The lifeboat crew endured forty-nine hours at sea to save the lives of eight men onboard the drifting lightship. The lifeboat involved, the *Mary Stanford*, is now at Ballycotton on permanent display above the harbour, having been rescued from rotting away in a dock.

Ballycotton is also well known as a sanctuary for American wading birds, as well as for many other species, blown across the Atlantic in autumn gales. I have been birdwatching here but never lucky enough to coincide with a 'fall' of migrants. There have been significant changes in the habitats here over the past century. These include an acceleration of coastal erosion, development of a coastal lagoon behind the shingle barrier in the 1930s, the subsequent loss of that lagoon in the early 1990s and its replacement by estuarine conditions. The loss of the lagoon led to a decline in both breeding and wintering ducks and swans but a corresponding increase in waders such as godwits, plovers and redshank.[20] Since then, some wintering species such as wigeon, teal, curlew and lapwing have declined, as they have in Ireland as a whole, but little egrets are now breeding there, and brent geese have increased. The complex of wetland sites between Ballycotton and Shanagarry is good for seeing waders at any time of year.

Cork Harbour

On a sailing trip along this coast, my friend Brian and I decided to make an overnight passage from Wexford to Cork. At first, the moon lit our path, but gathering clouds turned the route ahead pitch black. We each took the helm in turn for two-hour shifts while the other crew member grabbed some much-needed sleep. Steering into the night I was reliant on my charts and the flashing signals of distant lighthouses and navigation buoys. Occasionally, a ghostly seabird would fly alongside before disappearing into the gloom. As the dawn broke behind us we finally reached the sheltered waters of Cork Harbour and the 'welcoming' beam of Roche's Point Lighthouse. Here the meteorological station, just up the road from the lighthouse, records data of great importance to mariners. After passing through the narrow entrance between Roche's Point and Weaver's Point, we entered the comfortable marina at Crosshaven on the west side of the harbour. This involved some tricky manoeuvres due to the strong tidal currents passing through the pontoons, but we were eventually welcomed in the Royal Cork Yacht Club and invited to join an outdoor barbeque in the club grounds. This is the oldest yacht club in the world, which has been organising sailing here for over 300 years. It began on Haulbowline Island in 1720, where the headquarters of the Irish Naval Service is located today. The organisation of the club was then very much along the lines of the navy, with an admiral and captains. The rules show a clearly defined structure of sailing, with the admiral commanding the fleet and the other officers following in order of rank. Communication was by flags and gunfire, which enabled quite complicated manoeuvres to be completed. The gunpowder used by the race officers was mainly funded through fines, which were levied for such indiscretions as late arrival at the appointed rendezvous or for failing to attend meetings.

In Cork, there are several old, traditional boat designs that still sail today. In 1895, the Royal Cork Yacht Club commissioned a boat that could be replicated for competitive racing, could be crewed by four people, designed for the local conditions of Cork Harbour and built at that time for under £100 sterling. The first six boats were constructed and launched in 1896, with twelve Cork Harbour One-designs in total being built. In 1917 the United States joined World War I, with the first US Destroyers arriving in Cork, followed by submarines and a Flying Boat air base in Aghada in the eastern part of Cork Harbour. The following year Submarine Chasers, 110-foot wooden launches powered by gasoline engines, also arrived here. Their officers were mainly peacetime yachtsmen of the US Naval Reserve, and they were welcome visitors to the Royal Cork Yacht Club.

Not long before the outbreak of World War I the port of Cobh (then Queenstown) in the northern part of Cork Harbour became world famous for another reason. In 1912 the 'unsinkable' *Titanic* called to the port on her maiden voyage. She had set out from Southampton and called to Cherbourg before continuing on to Cobh, which was to be the last call before the transatlantic voyage to New York. A total of 123 passengers embarked here. Between passengers and crew, there were 2,206 people on board as she embarked on her final journey. Some 1,517 of these would never see New York, as she sank in the mid-Atlantic after colliding with an iceberg.

Just north of Crosshaven is the small inlet of Lough Beg, which dries out completely at low tide. The lough is surrounded by large industrial plants, several of which are run by multinational pharmaceutical companies.

These firms were among the first industries in Ireland to employ wind power to fuel their operations, and several huge wind turbines now dominate the landscape here. I was commissioned to study the possible impacts of these turbines on wintering birds on the neighbouring shorelines over a period of some years. This included the question of displacement, thought to arise from the movement and noise of the rotors, as well as possible collision by flying birds with the turbines. With several colleagues, I accumulated much evidence to show that there was no measurable impact and the birds continued to feed close to the turbines after they were built. Very few birds were recorded as direct collision victims, the most dramatic of which was a pelican that had escaped from Fota Wildlife Park across the harbour.

Praeger summed up the inner part of the harbour near Cork City thus: 'the seaward approach is erratic – a fifteen-mile zigzag channel through waters, sometimes broad, sometimes very narrow. The region between Cork and the ocean is then an archipelagic area, often richly wooded, with many villages and large residences – a picturesque and fertile region, very attractive, and quite unlike anything else in the country.'(WW) As we sailed into the northern part of Cork Harbour, known as Lough Mahon, I could see in the distance a Martello tower, partly ruined, on the edge of the railway that runs on a causeway past Marino Point onto Great Island. Having returned to tie up at the Cobh Sailing Club marina, I decided to visit the tower via the abandoned site of a fertiliser factory, and by walking along the edge of the railway embankment. My attention was drawn to the tower by the calls of a colony of terns nesting on top, a most unusual but nevertheless suitably secure site. This is one of a number of tern breeding sites around Cork Harbour, where the combined population reached a peak of 157 nests in 2015. They also nest on disused mooring structures in the Port of Cork at Ringaskiddy. Four of the tern chicks ringed in the harbour have been recovered on the west coast of Africa in Mauritania, Senegal and Togo, where they spend the winter months. This research has also produced evidence of interchange of terns from Cork with breeding colonies in Dublin, France and Britain.[21]

Harper's Island

One of the most sheltered parts of Cork Harbour

is at the northern edge around Harper's Island near Glounthane. Although there was an older hump-backed bridge leading to the island, it also became linked to the mainland in the 19th century by a causeway that still carries the Cork-to-Cobh local railway line. The island was probably fringed by saltmarsh until large embankments were built around its perimeter and the land was drained for agriculture. An abandoned farmhouse and lime kiln on the island testify to bygone farming days. I first visited Harper's Island in the 1990s, when it was still being intensively grazed. This was a time of economic growth signalled by the spread of motorways across Ireland. A route had been chosen for the new N25 linking Cork City with Midleton, and this cut right across the island from west to east. There was a period in the 1990s when flooding occurred regularly in winter and large flocks of waterbirds were attracted to the area. During the road construction work, the island was not being farmed and, presumably, the sluice was not being regularly maintained.

I went back for a visit to Harper's Island recently with Tom Gittings. He is a lifelong naturalist who grew up in central London and became involved, as a teenager, in the urban wildlife movement. Now there is no longer any farming at Harper's Island, and there is more water. The tide fills the lower parts and spreads out across former fields. The ponds contain both saline and fresh water, which offer a variety of feeding conditions for the birds. Two large shallow pools or 'scrapes' have been created, with a couple of small islands. These are covered with gravel to attract nesting terns that are short of other safe breeding places in Cork Harbour. The water levels in the 'scrapes' can be altered to attract feeding waders on the newly exposed mud, and the birds have responded. For the first time since this island was enclosed in the 19th century there were swirling flocks of ducks – wigeon, teal and waders – black-tailed godwit, golden plover and redshank. A solitary grey heron stood like a guard at the sluice, where the tidal water runs back out to the estuary. Nature seemed to be coming back to Harper's Island.

This impressive rewilding has been driven by a unique partnership. As well as undertaking the all-important bird surveys to monitor what is happening here, BirdWatch Ireland members Tom Gittings, Paul Moore and Jim Wilson worked closely with the local community in designing this facility and bringing the project to fruition. Conor O'Brien of Glounthaune Tidy Towns and Garry Tomlins of the local Men's Shed have been key members of the partnership. Two bird hides have been built at the end of screened walkways, and a third one is planned to allow visitors to get close-up views of the birds. A looped nature trail has been created around the island, and visions for the future include conversion of the old farmhouse to a visitor centre. In 2020 horses were introduced to graze the wet grassland and saltmarsh, but these are taken off in autumn when the main bird flocks arrive.

The proximity of Harper's Island to Cork City and the nearby railway station are both important for visitors using the site. The local National School resource teacher has produced education materials for school visits. The collaboration between local authority, national conservation organisation and the local community have been key factors in the success of the project. Between all the partners this has produced an exciting new habitat for birds and a welcome facility for both local and visiting birdwatchers.

Coastal wetlands like these may be only isolated fragments of a once-common habitat in the Irish landscape. However, managing them for wildlife is especially valuable for those species that cannot survive in the intensive agricultural landscape that has replaced most of the wetter habitats. If more such sites could be protected and restored, they could act as a network of stepping-stones for those birds and insects that need to migrate on a seasonal basis. The visitor facilities developed at these sites also give great opportunities for environmental education so that more people will understand the key importance of wetlands for nature.

Kinsale

After a short passage from Crosshaven in Cork Harbour we sailed south-west between the Sovereign Islands just off the inlet of Oysterhaven. From there the long outline of the Old Head of Kinsale came into view, with its distinctive lighthouse on the end. This was the location of the tragic loss of the passenger ship *Lusitania* in 1915. At the time this was the largest liner in the world, and most of the passengers were British people returning from America. Almost 1,200 people died in this sinking by a German U-boat, which contributed indirectly to the entry of the United States into World War I.

Sailing up the long, winding inlet that leads to Kinsale, we had a magnificent view of the surrounding beaches and hills. I imagined the fleet of twenty-six Spanish ships sailing here in 1601, carrying 3,500 soldiers who quickly took possession of the town. The subsequent siege of the town by English forces and the Battle of Kinsale placed this location at the centre of a series of events that were to have major implications for the English monarchy.[22]

On either side of the bay are the substantial battlements of Charles Fort and James Fort, built during the 17th century to protect this important harbour from invasion. Having moored the yacht in the marina of Kinsale Yacht Club, it was a short walk through the town and round to the east side to have a closer look at Charles Fort. First completed in 1682 and named after King Charles II, Charles Fort was sometimes referred to as the 'new fort' – to contrast with James Fort (the 'old fort'), which had been built on the other side of Kinsale Harbour before 1607. As one of the country's largest and best-preserved British military installations, Charles Fort has been part of some of the most momentous events of Irish history. During the Williamite Wars, for example, it withstood a thirteen-day siege before it finally fell. In the Civil War of the early 1920s, anti-Treaty forces on the retreat burned it out. This massive star-shaped structure was designed by the Surveyor-General Sir William Robinson, also the architect of the Royal Hospital Kilmainham in Dublin. Its dimensions were impressive, especially as I walked around the inside and outside of the walls, some of which are sixteen metres high. I tried to imagine the garrison of scarlet-uniformed soldiers manning the battlements and scanning the entrance to the harbour for any tall ships that would threaten the town. The houses within the fort are mainly roofless now, bare stone structures that would have had few comforts and probably held prisoners as well as the soldiers themselves.

The town of Kinsale is a very popular tourist destination, and is a great place to sample the best of Ireland's seafood. I decided to visit the well-known restaurant Fishy Fishy, which looks out from the quayside across the old harbour. Nearby, the reconstructed mast of a square-sailed ship gives a great impression of how the quaysides must have looked in previous centuries, when they were filled with vessels from all over the world. Here, I enjoyed a range of fish and shellfish including oysters, crabs and John Dory fish. The owner of the restaurant, Martin Shanahan, is one of Ireland's leading seafood chefs. In the 1980s he was a chef at the renowned Huntington Hotel, San Francisco, where his love of fish cookery intensified. After returning from the USA, Martin spent seven years as a fish retailer in Kinsale before opening a café. His passion and skill with fish became apparent, and this was brought to a wider audience through the television series *Martin's Mad About Fish*. He says, 'fish is nature's fast food. It is a gift, not a commodity, and deserves our respect'.[23]

West Cork

Leaving Kinsale on a single-handed passage, I rounded the lighthouse on the end of the Old Head and set sail along the south-west coast of Cork heading for the open Atlantic. Solo sailing is very different from having a crew on board, and a great way to learn good seamanship. My senses were constantly alert, watching for hazards like the floats of lobster pots which can get entangled in the propellor or rudder. I trimmed the sails frequently to get the best speed from the boat and adjust to changing wind speed and direction. I scanned the charts and the instruments regularly, estimating distances and travel times, monitoring wind speed, tide times and boat speed. Occasionally, I switched on the electronic steering and dropped into the cabin to make a cup of tea. Known as an autohelm, this is a vital accessory for any single-handed sailor. I didn't feel alone at all, as I knew the boat so well it felt like an old friend. The sun was shining and the sea was calm, so all was well with the world.

My first destination was the inlet of Glandore, which also holds the fishery harbour of Union Hall. The narrow entrance requires careful navigation, as two dangerous rocks, known as Adam and Eve, are right in the centre of the entrance channel. Seafarers are advised to 'avoid Adam and hug Eve' in order to stay in deep water. Once inside, the inlet is almost land-locked and the wooded shorelines make it one of the most attractive natural harbours in south-west Ireland. The sheltered waters allowed me to motor slowly into position and run along the deck to tie up to a mooring bouy in the bay. Then I rowed ashore to explore the area. After a hearty meal of seafood and a peaceful night on board I dropped the mooring and headed out to sea.

Just around Reen Point is another wonderfully sheltered inlet at Castlehaven, which was the scene of a fierce naval battle in 1601 between Spanish forces, who had landed in support of the Irish rebels, and the British Navy, led by Admiral Richard Levison. The Spanish had been welcomed here by the O'Driscoll family, who gave their allies possession of a castle on the north shore of the bay, the walls of which still stand today. This local engagement was eclipsed by the Battle of Kinsale, which put an end to Irish resistance, forced the Spaniards to withdraw and finally confirmed the English as rulers of Ireland, leading to centuries of repression. The Protestant ascendancy here in the 19th century was led by the Townshend family, who built the picturesque village of Castletownshend and populated it with people of their own denomination.[24] I walked down the steep main street to the edge of the bay, where an old corn mill dominates the shoreline. On the opposite side at Reen Pier, Colin Barnes moors his boat *Holly Jo*, in which he takes visitors out to see a variety of marine mammals on the West Cork coast.

Passing Toe Head on a flat, glassy sea surface, my yacht was joined by a school of common dolphins, riding in the small bow wave, breaching and playing underwater. Putting the steering on autohelm I moved up to the bow, where I was able to take close-up pictures of these lively mammals that are commonly attracted to boats. This area is one of the best places in Ireland to see whales and dolphins. Whale-watching boats leave each day in summer from Castlehaven and Baltimore. From April to December it is possible to see minke whales, common dolphins, bottlenose dolphins and harbour porpoises. In late summer and autumn they are often joined by humpback and fin whales, which come inshore to feed. They can sometimes be watched from vantage points on shore as well as from boats. Other marine animals regularly seen on these coasts include the ocean sunfish and the basking shark, the largest fish in the Atlantic Ocean. Leatherback turtles and bluefin tuna are often seen around here too, and are increasing in numbers with warming of the seawater due to climate change.

I steered a wide course around some jagged rocks here known as the Stags. In the 1980s, the threat of oil pollution was a very real issue, as many licences for offshore exploration were being issued by the Irish government. In November 1986, our worst fears were realised when the *Kowloon Bridge*, a bulk carrier with a cargo of iron ore pellets, lost power off the West Cork coast, drifted east and ran aground on the Stags Rocks. The resulting fuel spill spread out along the south coast, causing extensive damage to local wildlife and fisheries. In the six months that followed, the oil seeped into hundreds of inlets, beaches and rocky shores with varying degrees of damage. This coincided with the return of seabirds to their breeding colonies, and volunteers walking the beaches recovered hundreds of oil-damaged victims, most of which were already dead.

Lough Hyne

Having tied up outside the harbour at Baltimore, I decided to explore a unique habitat nearby which has the distinction of hosting the longest continuous series of marine biological studies in Ireland. Lough Hyne is a land-locked saltwater inlet, with its only connection to the sea being a narrow channel through which the tide runs at a fast pace over a series of rapids. I climbed the steep, rocky peninsula to the west and looked down on the rapids, which are like a fast-flowing river at low tide. The naturalist and filmmaker Gerrit van Gelderen wrote a vivid description of this unique feature:

> The tide rises for about four hours but falls for about eight hours. This means that, even after the Atlantic tide starts to rise again the water level in the lake is still dropping until the very moment when the two meet. Then, for less than a minute, the current comes to a complete standstill. A hush falls over the waters, an unnatural quiet, and the weeds that have been pointing seaward quiver, stand up and when the reversing tide has gathered momentum, fall back again pointing inland. It is a moment when sea and lake seem to hold their breaths and, if you were there waiting, you would never forget it.[25]

Some friends of mine had recently sailed two small boats through the channel at high tide, making a speedy exit while there was still sufficient depth. The inlet inside is deep and mysterious, and surrounded by rocky shores.

In 1886 Lough Hyne was 'discovered' by marine biologists, but their studies were interrupted by the War of Independence in Ireland. In 1922, Louis Renouf, who was born on the Channel Islands near France, was appointed as Professor of Biology at University

College Cork. Praeger met Renouf at an early date and told him of the wonders of Lough Hyne, encouraging him to study it. Praeger was clearly impressed by Lough Hyne, writing that,

> it resembles a gigantic marine aquarium, and the peculiar conditions of life have remarkable repercussions on the flora and fauna. Many forms common on the surrounding shores are absent or very rare in the lough ... many animals and some plants grow to quite unusual dimensions. The fauna is characterised by a remarkable abundance and variety of species, particularly of sedentary and sessile species. ... To lean over the side of a boat and view the bottom with a 'water-telescope' is like peeping into a strange calcareous sort of fairyland.(WW)

In 1925 Renouf set up a marine laboratory here and built a wooden hut by the water's edge. This was later adopted by the University, allowing generations of biology students to undertake their fieldwork here. Trevor Norton, later Professor of Marine Biology at the University of Liverpool, began his academic career in

1964 as a student who was inspired by the wonders of Lough Hyne, so much so that he wrote a lighthearted book about his experiences there. He described the book as 'a biography of the lough, an inlet of the sea masquerading as a lake, and one of the most renowned ecological sites in the world'.[26]

A number of distinct habitats have been identified in Lough Hyne, including the intertidal zone between high and low water mark; shallow rocky areas down to about two metres below low water mark; fine mud zones covered with broken mollusc shells and red seaweeds; the mud-burrow zone with animals such as the Dublin Bay prawn; and the very deepest parts that have worm tubes standing up out of the mud.[27] The great variety of marine habitats and water movements within the lough support several thousand species of marine plants and animals, making a unique assemblage in such a small area. This includes over 70 species of sponges, twenty-four species of crabs, eighteen species of sea anemones as well as three-quarters of the species of sea slug known in Ireland, and almost one-third of all the seaweeds recorded in this country. Many of the species are better known from southern Europe and the Mediterranean, suggesting that the habitats and water temperature here are particularly suitable for them. Unfortunately, there is strong evidence of a deterioration in water quality in the lough, with consequent changes in biodiversity. The causes are thought to be increased use of the area as a tourist site, arrival of invasive and non-native species and climate change warming the seawater.[28]

The lake is also popular for watersports. On a summer evening I was invited to join a kayaking outing here. This trip was a unique experience, being on the water from dusk into darkness. In the darkening sky I was mesmerised by the sounds of wading birds coming into roost, the sunset followed by a rising moon, the aromas of honeysuckle and gorse on the breeze, the panoply of stars overhead, the astonishing bioluminescence on the water and the deep, dark peace and serenity of night. It was a special coastal trip that I will remember for a long time.

Roaringwater Bay

The popular tourist destination of Baltimore turned out to be a good base to explore the islands of Roaringwater Bay, and I was joined here by a friend, Mark, and my son Derry. Praeger's memories of Roaringwater Bay give a picture of West Cork around the beginning of the 20th century:

> A meandering railway penetrates to Schull, and the roads are as good as you could expect them to be in so lonely a country. All is furzy heath and rocky knolls, little fields and white cottages and illimitable sea, foam-rimmed where it meets the land, its horizon broken only by the fantastic fragment of rock crowned by a tall lighthouse which is the famous Fastnet.(WW)

Our berth was on a small pontoon outside the harbour at Baltimore, while towering over the village is the well-preserved Baltimore Castle, built in 1215 by the Anglo-Norman Lord Sleynie. It later became the seat of the powerful O'Driscoll clan, and was used as a base for their trading activities and pirate ships. From here they conducted an ongoing rivalry and warfare with the ruling clans of Waterford. In 1631, Baltimore was attacked by Algerian pirates, who took over a hundred local people captive and carried them off to be sold as slaves in north Africa. In 1649 the castle fell into the hands of Cromwell's troops, who set up a garrison here.[29] For centuries afterwards it was a ruin but, following extensive restoration by the owners, Patrick and Bernie McCarthy, it now contains an exhibition of its pirate history and archaeological finds. I chatted to them for a while and discovered that they actually live in the castle as well as opening it to visitors. A walk on the battlements gave me fine views of Roaringwater Bay and the offshore islands.

Out to the south-west of Baltimore lie the two largest islands, Sherkin and Cape Clear, both of which have permanent populations. On Sherkin Island, we tied up alongside some pontoons, which are moored just below the hotel and provide ideal shelter from westerly winds. From here it was a short walk across to Kinish Harbour, a wide sandy bay that splits the island in two. To the north-west end is Sherkin Island Marine Station, a unique family-run study centre that has been operating here since it was established in 1975. I met the founder, Matt Murphy, who took me on a tour of the station. This now comprises a large complex of five laboratories and a library, together with an herbarium of plants and seaweeds. Biologists at the centre, and other enthusiastic volunteers from many countries, have undertaken long-term monitoring of marine life in Roaringwater Bay, as well as in Bantry Bay and Cork Harbour. We finished the day with a healthy bowl of fresh mussels in the hotel above the pier.

The next morning was fine, with a light north-westerly breeze, ideal conditions for a trip out to the iconic Fastnet Lighthouse. Sailing along the rugged south coast of Sherkin we passed close to a minke whale feeding lazily on the surface, while gannets dived into the waters all around. As always, the Fastnet Rock was surrounded by a good Atlantic swell, and landing was nigh impossible. This was often the last part of Ireland seen by departing emigrants bound for America in the 19th century, and it thus gained the nickname 'Ireland's Teardrop'. The building of the tower, started in 1899, was captured in a wonderful set of glass-plate photographs by the Commissioners of Irish Lights.[30] The intricate masonry work comprises eighty-nine courses or layers of cut stone made up of 2,074 blocks. The lower part of the tower is well below sea level. Richard Taylor, who spent over forty years servicing lighthouses around Ireland, wrote that this was one of the few sites where the keepers had to live in the tower itself. 'For exercise during those enforced sojurns, endless journeys would be made across the service-room floor, each man going in opposite directions.'[31] In 1979 the famous Fastnet Yacht Race, which departs from the Solent in southern England, met a massive storm here, and many yachts were wrecked, with a total of fifteen deaths.[32]

With the wind in the north-west we had a lively sail back to the North Harbour of Cape Clear Island. This is the most southerly populated part of Ireland, and farming continues on the island in one of the most exposed locations in the country. Its position at the end of a long, partly submerged peninsula makes it a classic location for bird migration in and out of Ireland. An original account of the island's natural history was published by Tim Sharrock, one of the founders in 1959 of the longest-running bird observatory in the Republic of Ireland.[33] This was updated in 2020 with a new book by the warden of the observatory, Steve Wing.[34] Having visited the observatory building on the quayside, we walked across the steep, narrow roads to the cliff-bound South Harbour and then east along the ridge of the main island. There are a few vehicles on the island, and it is wise to stand well back while they pass, as many are salvaged from the mainland, having failed their road tests. A return walk along the southern cliffs gave fine views of passing seabirds which are one of the attractions of this island for birdwatchers.

Back on the yacht, we sailed along the north side of Cape Clear and close to the eastern end of the three Calf Islands. On the way, we passed Heir Island, which lies just a few hundred metres off the mainland. A short ferry ride from Cunnamore Quay brings visitors and householders into the island. At one time this was a busy boat-building centre, when the lobster boats of Heir Island and Roaringwater Bay were once among the best known and most distinctive fishing boats of the region. This unique fleet set out under sail and oar each year to fish lobsters and crayfish all along the coasts of Cork and Waterford. English fish merchants were sending boats from Southampton to the Irish south coast to buy lobsters and crayfish and, by 1892, there were forty lobster boats registered in the Congested District of Baltimore with the majority of these based out of Heir Island. These were gaff-rigged sailing boats with a crew of three. The sails were made by the fishermen themselves from heavy calico or cotton, and they were 'barked' to preserve them. Tree bark was boiled in water and the resulting solution gave the sails their distinctive tan colour.

Peter Somerville-Large made a cycle tour of the entire coast of west Cork in the early 1970s. He recounted how the smaller islands in Roaringwater Bay had been deserted, one by one. He described how he had visited Horse Island opposite Ballydehob just a few years earlier. 'The last people there, an elderly couple, were living all alone. Next year they were gone. The house, still intact and comfortable, stood empty, the linoleum in place, last year's calendar on the wall. Down by the pier a plough had been thrown into the water where it looked like a gesture of despair.'[35]

From here our next stop was the large natural harbour of Schull, which is well protected from southerly winds by Long Island across the harbour mouth. Here we picked up a mooring for the night. I visited Schull Community College, which must be one of the few secondary schools with its own slip running straight into the sea. The college has an on-site sailing centre called the Fastnet Marine Outdoor Education Centre. This offers a range of maritime courses as well as regular sailing for students in the school's own fleet of dinghies. Schull sailors have represented Ireland at individual and team international events. The village street in Schull is full of restaurants, fashion and craft shops, highlighting its popularity as a holiday resort for people from Cork City and further afield. Standing outside one of the pubs with creamy pints we listened to maritime stories from seafarers of all kinds.

Leaving Schull, we sailed along Long Island Channel, past Toormore Bay and into the sheltered inlet of Crookhaven, which was the last port of call on my yacht trip along the south coast. This deep natural harbour opens to the north-east, so is very sheltered from the Atlantic and a popular anchorage for cruising yachts. The village is busy with tourists in summer, so we joined them and enjoyed a good lunch of seafood chowder and locally caught crab in the sunshine outside O'Sullivan's Bar. As the most south-westerly harbour in Ireland, this has been a strategic location for shipping over the centuries. During the Napoleonic Wars a provisioning depot was established here so that visiting warships could be supplied with wheat, oats, corn and butter. In the mid-19th century mail from Ireland was collected here by ships as their last stop on the way to America. In the early 20th century, when herring were abundant on the coasts of Ireland, a fishing fleet based in the Isle of Man would anchor here between March and June each year to exploit the passing shoals. Here too are the remains of a pilchard palace, once owned by the Earl of Cork and Sir William Hull. These fish-processing stations once flourished on the shores of Roaringwater Bay and around much of the west coast of Ireland. They probably reached their heyday in the 17th century, when pilchards were the main catch. Huge shoals of these small fish came to the comparatively warm, sheltered waters of the islands during the summer months, along with other oily fish such as herring and mackerel. The buildings contained iron presses to squeeze the oil from the fish before they were preserved in salt. Today, pilchards are rare in Irish waters due to a combination of overfishing and changing climate.

Having explored the inlet of Crookhaven and the dunes at Barley Cove, I walked out to the lighthouse at Mizen Head. Accompanied by a pair of choughs, I ventured onto the awe-inspiring metal suspension bridge across a deep chasm, from which dramatic geological formations of Old Red Sandstone were clearly visible. Mizen Head Fog Signal Station was sanctioned in 1906 by the Irish Lights Board to combat the high loss of life and shipping on the rocks below. The Station was manned by three keepers until 1993, when the lighthouse was automated. A community rural development initiative was then formed to create the tourist centre, 'Mizen Vision', in the former keepers' quarters, while taking a lease on the path to the lighthouse from the Commissioners of Irish Lights.

I walked out to the next promontory north of Mizen, where the remains of old castle walls still stand on the edge of the cliffs at Three Castle Head. This coastline has been infamously associated with smuggling and piracy for many centuries. As recently as 2007 this was the scene of a dramatic discovery of cocaine, washed up on the coast here and valued at €440 million. The attempt to land the illegal drugs failed when one of the men filled their petrol-powered outboard engine with diesel by mistake. The inflatable launch overturned and dumped sixty-two bales of cocaine into the sea. Three of the men involved in the operation were sentenced in 2008 for a total of eighty-five years. This headland is where the coast turns north, and the great flooded inlets of the south-west begin.

Leaving the South Coast

The south coast of Ireland is fringed by the Celtic Sea, which stretches away to merge with the English

Channel and the Bay of Biscay. But it also feels the strong influence of the Atlantic Ocean coming in from the west. Being closest to the European continent, the south coast has the warmest climate in Ireland but, occasionally, great storms and rainfall wash over it. This coast is frequently the first landfall of migrant birds moving north from Africa, and it has the distinction of being the best place in Ireland to watch large whales and dolphins in a natural setting. The human history of the south coast is dominated by trade with Europe, and its large natural harbours, such as Waterford and Cork, are havens for shipping in stormy conditions. The sailing waters of West Cork offer unrivalled cruising grounds, with numerous islands and welcoming harbours. I ended my sailing trip in the south-west because cruising in the Atlantic is more risky, due to big seas and a relative lack of sheltered harbours. My exploration of the remainder of the Irish coast was mainly by land, except where a boat or an aircraft was needed to reach remote islands.

West Coast

By the end of the 19th century, Praeger was clearly exhausted after the massive effort to complete his huge work, *Irish Topographical Botany.* He had been lodging with his friends the Tatlows near Dundrum, County Dublin, and would have walked or cycled from there to and from the National Library each day. Mrs Tatlow was a naturalist interested in marine molluscs, and about this time she and her husband left Ireland to live in Germany. In the autumn of 1901 Praeger, now aged 36, was persuaded to visit them to help him relax after his intense work in the previous five years. It was there he met a young woman named Hedwig Magnussen and, following a whirlwind romance, they were married the following year. A letter from one of their Dublin friends noted:

> Dr. and Mrs. Praeger were always so happy together that it was a pleasure to be with them. Each, in turn, told me the story of their engagement. Praeger said that he had 'no thoughts of matrimony in his mind' as he started off on his holiday and met Hedwig, a friend of the Tatlows. He could not speak German and she could only speak what he called 'schoolgirl English' (she had been for a while at an English school). Yet within the fortnight they managed to get engaged. From the first she learned to share his interest in botany and also to share his love for the beautiful scenery of the coasts of Ireland.'[1]

In the early 20th century, due to Praeger's extensive experience of all Irish counties, he was invited to write guides to different parts of the country for tourists travelling on the new railway lines. This was followed by publication of a whole series of 'Official Guides' for routes such as those managed by the Belfast and County Down Railway. In his autobiographical book *The Way that I Went* he often slipped easily into the role of a tour guide to the landscape. 'If you ask what is the best county in Ireland to walk in, I reply Donegal;

the best region to cycle in, Connemara and its natural adjunct West Mayo, or alternatively Kerry.'

The successful collaboration organised by Praeger for the Lambay Survey of 1905–06 now appears like a trial run for the much larger Clare Island Survey that was to follow. The Mayo island was chosen by Praeger because it offered a large enough unit, well separated from the mainland, and could also provide a small hotel as a base for the dozens of naturalists that he planned to attract. He was absorbed by the question of how species colonised isolated areas of land and their methods of dispersal. It should be remembered that this was just a few decades after the publication of Darwin's landmark book *On the Origin of Species*, and the biological community was puzzling over how different races of plants and animals, isolated from each other over long periods of time, could evolve into separate species. Darwin had demonstrated how small oceanic islands, such as the Galapagos off South America, can have a whole suite of species that are not found anywhere else, and naturalists in Ireland were searching for any endemic Irish species to show that nature in Ireland was equally special. Praeger was especially interested in how the flora of Ireland differed from that of the neighbouring island of Britain, and from the near continent of Europe. This movement was also happening at a time of great cultural and nationalist revival, when Irish art, literature and language were coming more into focus. The concluding remarks in *The Way that I Went* capture the inspiration that Praeger found in the west coast and its outstanding value for tourism: 'The Atlantic fringe, with its tall brown hills, its tattered coast-line and its snowy foam, is the region to which one's errant thoughts recurrently stray, and which remains a lodestar to people of many lands – a magic region which, once viewed by the stranger, rests forever in his memory.' Today, the west coast is branded for tourism as the Wild Atlantic Way, forming one of the longest defined coastal routes in the world. It stretches all the way from West Cork to north Donegal, following a tangled trail of coastal roadways. It was along this route that I set out to explore some wilder areas, mainly on foot.

Sheep's Head

Lying between Dunmanus and Bantry bays, the long Sheep's Head Peninsula is the smallest of the five narrow ridges of land that jut out into the Atlantic

Ocean. Although remote by road, the Sheep's Head Way was one of the first long-distance walking routes in the country. It is a 200-kilometre waymarked route that follows the perimeter of the peninsula and includes eighteen looped walks for shorter outings. It includes some fine cliff coasts, a ruined village and old copper mines. Some of the old cottage ruins on this route, abandoned in the 1940s, are known as the 'Crimea'. This strange name arose as a result of constant arguments among the families who lived here after the Great Famine. This prompted a visitor to liken it to the Crimean War, which was taking place at the time.[2] On a fine winter's day, I followed one of the looped walks on the south side of the peninsula. This started at the small village of Ahakista, where I strolled along a quiet country road turning south on the coast around Farranamanagh Lough and ending at the next village of Kilcrohane. Trumpeting calls coming from the lake alerted me to the presence of a small flock of whooper swans, winter visitors from Iceland. The outlet from the lake flows beneath an impressive stone clapper bridge. The path also passes by some ruined buildings which were once a Bardic School. Bards (or *file*) were highly respected people in Irish clans, as they were part poets, part prophets who were expected to know the history of their people and give them insights into the future. Young men were sent here to be trained for at least seven years. When qualified, a bard was assigned to a particular clan chief, where he was responsible for continuing the spoken traditions of the clan.[3]

Just above the village of Kilcrohane lies a farm that was home to Jack Sheehan from 1920 to 2003. This ordinary man's long life paralleled the development of the Irish state. He was one of eleven children born into an impoverished family and grew up in hungry times, when emigration offered the only prospects for most of his siblings. He took over the farm when his father died and cared for the fields throughout his eighty-three years. He lived alone here through the period when all farming depended on manual labour and horsepower, leading into the phase of mechanisation and rural electrification and finally to farming supports from the European Union. For Jack this was home, and he rarely left it except to bring his milk to the creamery.[4] I felt that the story of Jack and his farm was probably representative of many smallholders in the west of Ireland over the last century.

At Ahakista, I went to visit the Heron Gallery, which showcases the work of artist Annabel Langrish. Her compelling paintings focus on the diversity of wildlife that surrounds her home here on the Sheep's Head Peninsula. The gardens around the Gallery, designed with an artist's eye, were a great spot to enjoy the gorgeous, wholesome local food I found in the lovely café. The organic gardens are informal, with colourful flowerbeds and a walk around a large pond surrounded by wildflower meadows and young woodland. There is also an orchard of Irish heritage apples, two polytunnels and raised vegetable beds growing all the salads and vegetables used in the café. I followed a path leading up to the ridge, where there is a great view down to Dunmanus Bay.

Bantry Bay

When he was a young man in his twenties, Praeger was selected to join a pioneering expedition in 1888 to study marine life in the deep waters off the West Cork coast. In common with the north-western Rockall Trough, this is one of the areas where the continental shelf at the edge of Europe comes closest to the Irish coastline. The leader of the expedition was a remarkable man with the delightful name of William Spotswood Green. He was thirteen years older than Praeger and was, at the time, a Church of Ireland rector in the parish of Carrigaline on the west side of Cork Harbour. But Green's real love was the sea and, when his church duties allowed, he spent his time exploring the seashore around the harbour and studying fish and marine life.

Earlier in 1888 Green had presented a report 'On the Fisheries of the South and South-West Coast of Ireland' to the Royal Dublin Society (RDS), in which he highlighted the quality of the Irish spring mackerel and the high returns that could be achieved. He wrote that the mackerel were 'very large, from two and a half to three pounds each and fetch nearly twice the price in Billingsgate of the mackerel caught on the English coast'. Later that year the Royal Irish Academy agreed to fund one of his deep-sea research cruises in a ship called *Flying Falcon*. After some initial trawling in Bantry Bay, West Cork, the expedition turned west and sank the dredge in 345 fathoms (631 metres) of water almost 100 kilometres from land. Praeger wrote that 'the introduction of steel-wire rope for dredging and thin steel wire for sounding, had made these operations much simpler than when hemp had been employed'. (WW)

The next dredge reached down to almost 2,000 metres to explore a then unknown world of the deep sea, which was in total darkness. A further sample from over 2,300 metres took an hour to recover from the seabed and 'in its net a variety of strange and beautiful animals', which must have appeared exciting to these Victorian naturalists. After surviving a gale at sea, the scientists made a final haul which contained, 'a splendid catch of marvellous creatures. There were great sea-slugs, red, purple and green; beautiful corals, numerous sea urchins with long slender spines, a great variety of starfishes of many shapes and of all colours including one like a raw beefsteak, which belonged to a new genus; strange fishes and many other forms of life.' Battered by further storms the ship turned for Bearhaven in Bantry Bay. Praeger wrote, 'I remember that when at last grey dawn gladdened our sleepless eyes, Green's face, ghastly white under a sou'-wester, appeared at the cabin door and looked around the wreckage strewn on the floor.' In 1890 Green left his parish and became Inspector of Fisheries for the RDS. Two years later he was appointed to the

Congested Districts Board, which had been set up by the government to try to relieve poverty and improve the prospects of the people on the western seaboard. It focused on the potential of fishing by building local piers and supplying new boats and nets.

Castletown Bearhaven, near the entrance to Bantry Bay, is today a thriving fishing port. Opposite the harbour I walked right around the island of Dinish, which has been developed as a fish-processing centre and is cluttered with the modern paraphernalia of fishing – nets, winches and boats under repair. Towards the head of Bantry Bay is the village of Glengarriff and another island known as Garnish (or Ilnacullin), which contains a unique garden with many rare plants that survive here because of the frost-free conditions. Travelling out by boat to the island I was fascinated by the groups of harbour seals that lay on rocks apparently undisturbed by local ferries that pass close to them several times each day. One tree on the island was chosen in 2014 by a pair of recently released white-tailed eagles to build their large, bulky nest, but they only managed to successfully rear chicks in a few of the subsequent years. In the summer of 2020, I was transfixed by live video footage of these magnificent birds as they ferried fish and seabirds that they had caught in the bay, back to a single growing chick in the nest. These eagles were once a relatively common sight on the west coast, but they were gradually exterminated by shooting, poisoning and egg collecting, until the last pair nested in County Mayo in 1912. The reintroduction of these eagles to Ireland began between 2007 and 2011, when a hundred young white-tailed eagles were brought from Norway and released at Killarney National Park. They have since spread throughout the west of Ireland, with up to fourteen breeding pairs formed in some years, most of these on inland lakes. The pair breeding in Bantry Bay has the only known coastal nest site in Ireland.[5]

Just off the town of Bantry is Whiddy Island, which had a population of over 500 in the mid-19th century. The pilchard industry was the main source of income for the islanders, who processed the fish for export as they came ashore. The island remained relatively unknown until the late 1960s, when a large oil terminal was constructed here. This was designed to accommodate the largest supertankers sailing directly from the Middle East. In 1979, a French tanker, the *Betelgeuse*, exploded while it was unloading a cargo of crude oil at the terminal. The blast and subsequent fire killed fifty people, while the oil continued to leak into the bay for weeks afterwards. This is still considered to be one of the worst maritime disasters in Irish history. The terminal was never fully repaired but was transferred to the Irish government in 1986 and used to hold the Irish strategic oil reserve. The long, semi-derelict jetty has now become a novel artificial nest site for breeding seabirds, with fifty-three individual black guillemots found there during a pre-breeding survey in 2017. There are also up to ninety shag nests and plenty of gulls breeding on it.

While visiting Bantry, I managed to attend a lecture about the life of an exceptional local scientist. Ellen Hutchins was born in 1785 at Ballylickey, near Bantry. One of twenty-one children, of whom only six survived to adulthood, she later became a cryptogamic botanist, who collected and catalogued over 1,100 species of plants in her area. She was generous with her knowledge, and sent preserved specimens as well as drawings of plants to leading botanists throughout England and Ireland. She found around twenty seaweeds, lichens, mosses and liverworts that were 'new to science', while a number of her rarer finds –

three species of lichens, two species of seaweed, and a moss, Hutchins' pincushion – were named in her honour. Her specimens and records are preserved at the Royal Botanic Gardens in Kew, London; Trinity College Dublin; the Natural History Museum, London; and the New York Botanical Garden, where they are still used for research. Ellen Hutchins never published on her own behalf, but her finds, her observations and praise for her as a botanist appeared in botanical publications of the day. Ellen suffered from poor health, and she died in 1815, just before her 30th birthday. In her short life, she had excelled in botanical discovery in an era long before women were accepted in the sciences. Praeger called her 'a botanist of great promise'[6] and, had she enjoyed a longer life, she might indeed have rivalled his own botanical prowess. The annual Ellen Hutchins Festival is held in her honour each year at Bantry.

Skelligs

At the end of the Beara Peninsula, the aerial cable car that crosses high above rough waters to Dursey Island is a spectacular feature. From the clifftop here I could see the distinctive triangular outlines of the Skelligs to the north. Praeger wrote that these islands were 'one of the few places in Ireland, to my sorrow, I have not succeeded in reaching. Thrice I waited about Valentia for a good day, but the sea continued to run high, and landing was declared impossible, though it was midsummer.' Then, in a philosophical mood, he added, 'I suppose it is salutary that some of one's desires in this world should remain unfulfilled, even if they are as modest as this one.'(WW)

I was more fortunate to arrive in good weather. Joining a ferry from the shelter of Portmagee at the western end of Valentia Island, we headed south-west towards the two Skellig Rocks lying fourteen kilometres away. The sea surface was relatively calm, although there is always some swell as the ocean rolls in from North America. As our boat made its way offshore, we were accompanied by a school of common dolphins, bow-riding and playing in the wake. All around us there were seabirds – puffins, guillemots, razorbills, shags and increasing numbers of gannets – diving vertically into the sea. With the boat finally tied up at the old stone-built pier in Blindman's Cove, the larger of the two islands, Skellig Michael, towered above us in a menacing way. Climbing ashore involved a tense wait until the boat reached the top of the swell and then a quick jump before the deck dropped again several metres. I was on a much-anticipated family visit to this iconic island and, with two small boys on board, we had to take extra care, especially on the long climb up the stone steps, with a two-year-old on my shoulders.

The boys, Rowan and Derry, were fascinated by the tameness of some of the birds, with puffins standing just a few metres from the path and some of them even nesting beneath the steps. This island also holds a sizeable population of storm petrels nesting in crevasses in the old stone walls, the only evidence of their presence during the day being a churring sound that I could hear as I walked along the path to the lighthouse. The climb to the monastery site, at over 180 metres above sea level, was rewarded with a wonderful view of the west Kerry coastline and the heaving Atlantic far below. It is hard to believe that a small band of early Christian monks built this complex of tiny beehive huts entirely by hand without the aid of any modern machinery and lived here in such an isolated spot far from civilisation. They must have subsisted by collecting rainwater and harvesting seabirds and their eggs, but I wonder how they survived in winter, when the islands could be cut off from help for many weeks or months.

The playwright and author George Bernard Shaw wrote a letter to a friend in 1910 in which he expressed his amazement following a visit to the Skelligs:

> Yesterday, I left the Kerry coast in an open boat, 33 feet long, propelled by ten men on five oars. These men started on 49 strokes a minute, a rate which I did not believe they could keep up for five minutes. They kept it without slackening half a second for two hours, at the end of which they landed me on the most fantastic and impossible rock in the world: Skellig Michael, or the Great Skellig, where in south west gales the spray knocks stones out of the lighthouse keeper's house, 160 feet above calm sea level ... An incredible, impossible, mad place, which still tempts devotees to make 'stations' of every stair landing, and to creep through 'Needle's eyes' at impossible altitudes, and kiss 'stones of pain' jutting out 700 feet above the Atlantic. ... I tell you the thing does not belong to any world that you and I have lived and worked in: it is part of our dream world.

The smaller neighbouring island of Little Skellig is even more precipitous than Skellig Michael, and there is no safe landing place. Here is Ireland's most numerous breeding colony of gannets, the largest seabird in the North Atlantic. Gannets spend most of their lives on the open sea and feed by diving on small shoaling fish such as sandeels as well as mackerel, herring and discarded fish waste from fishing vessels. The most recent estimate of the size of this colony came to over 35,000 nest sites, making up about three-quarters of the Irish gannet population.[7] Circling the rock in the boat, I was in awe of the sheer mass of birds nesting on every ledge and space available, the whirring of wings and loud calls, but most of all by the pungent smell of guano, or bird droppings. The fat young gannets in late summer were a valued resource for coastal communities in previous centuries, leading to serious competition between the people of the Blasket Islands and neighbouring coasts. Tomás O'Crohan, in his famous book *The Islandman*, tells of an expedition to collect the birds for winter food:

> A boat with a crew of twelve men used to be guarding the rock, well paid by the man who

owned it. This time a boat set sail from Dunquin, my father among them, and they never rested until they got to the rock at daybreak. They sprang to it and fell to gathering the birds into the boat at full speed and it was easy to collect a load of them, for every single one of those young birds was as heavy as a fat goose.[8] Des Lavelle, the boatman on our visit, told me that as late as 1869, the Little Skellig was rented annually for the taking of feathers and young gannets, many of which were sold to the local population. Now retired, Des is an accomplished photographer and the author of an excellent book about the Skelligs.[9]

Valentia Island

As most of the ferries for the Skelligs leave from Portmagee, it made sense for us to stay overnight on nearby Valentia Island, so we booked into the comfortable guest house run by Des and Pat Lavelle at that time. This is one of a line of houses named Cable Terrace because they were built at the time of the first commercial transatlantic cable that was laid from here to Newfoundland in the mid-19th century. Prior to this, two weeks was the fastest time that a message could be delivered to North America from Europe, as all communications were then sent by ship. The first attempt to lay the cable in 1857 was a failure, when the line snapped just 380 nautical miles from Valentia, and could not be recovered from the seabed. A number of other unsuccessful attempts were made, but it took several years before the final cable was pulled ashore at a tiny fishing village in Newfoundland, some 1,686 nautical miles from Valentia Island. To mark the new communication link, Queen Victoria sent a message to the President of the United States at the time, James Buchanan: 'The Queen congratulates the President on the successful completion of an undertaking which she hopes may serve as an additional bond of Union between the United States and England.' This cable served its purpose for some eighty years, until the first satellites changed global communications forever and the Transatlantic Cable Station on Valentia closed its doors for the last time in 1966.

Valentia was also the home of a famous marine biologist, Maude Delap, who was born in 1866, just one year after Praeger. Maude and her sisters were fascinated by the diverse marine life on the shores of the island, so they sent specimens to the Natural History Museum in Dublin for identification. This led a team of leading biologists of the day to undertake a survey of the island in 1895 under the auspices of the Royal Irish Academy. Maude also conducted her own experiments on rearing jellyfish, and she was the first person to get them to breed successfully in captivity. She wrote a series of influential articles, all under her own name, which was quite unusual for a woman at the time. She once discovered an extremely rare True's beaked whale washed up on the shore of Valentia. She buried the carcass in her garden and dug it up again a few years later when asked for the skeleton by the Natural History Museum. Due to her significant contributions to marine biology, Maude was offered a position in the Marine Biological Station at Plymouth in 1906, but her father, the local rector, refused to let her go without marrying first. However, she remained single and spent her entire life on the island, where she continued to collect samples and study them in her home laboratory. Even in old age she continued her scientific work, and often went out fishing in a small rowing boat.

While I was staying on Valentia I went to visit the old slate mines on the north side of the island. Entering one of the mine openings I was greeted by a family of choughs that were nesting on a ledge, high on a rock wall. The parents shrieked their warnings and the chicks froze, hoping they would not be noticed. Valentia Slate Quarry was first opened in 1816 to supply slates, mainly for roofing and flooring. Valentia Slate was used extensively in many famous buildings in London, including the Houses of Parliament, Westminster Abbey and Cathedral, St Paul's Cathedral and many of the stations on the London Underground, such as Waterloo, Charing Cross, Liverpool Street and Blackfriars. It was also used in the Paris Opera House. During its best years in the 1850s the quarry employed up to 500 workers, but competition from cheaper and softer Welsh slate forced the historic quarry to close in 1911. In recent years, a new business has been established in the slate quarry, producing a wide range of quality products including floor tiles, countertops, windowsills, fireplaces, bath surrounds, garden furniture, stairs and grave memorials. The rocks near the slate quarry were the location for the remarkable find in the 1990s of a series of large footprints of an ancient creature known as a tetrapod. At least eight separate trackways were found here, with the longest being fifteen metres in length. Thought to have been made around 360 million years ago by a four-legged amphibian up to a metre in length, the tracks provide the oldest reliably dated evidence of four-legged animals moving over land.[10]

The fascinating geology and archaeology of Valentia were studied in great detail by Professor Frank Mitchell of Trinity College Dublin. The bedrock here, as in much of south-west Ireland, is Old Red Sandstone, but this comes in both coarse and fine versions, the latter forming the slate. In his autobiography, Mitchell wrote:

> I know Valentia well because I worked there for twelve years – often staying a month at a time – chiefly trying to trace the archaeological features that were swallowed up by the growth of peat and later partly revealed by turf-cutting over a long period of time. The island is extraordinarily rich in antiquities, which stretch from the Mesolithic peoples about 6,500 years ago to the industrial archaeology, quarries, cable-stations and lighthouses of the past century.[11]

In his scholarly book, *Man and Environment in Valentia Island*, Mitchell charted in great detail the ebb and flow of human activity that he interpreted from heaps of stones in the fields and bogs. Pollen analysis from remains in the peat also allowed him to unravel the vegetation history from the ancient woodlands to the farming practices of the 20th century.[12]

Castlemaine Harbour

The long sand dune points of Rossbeigh and Inch act like baffles at the mouth of Castlemaine Harbour, which lies between the Iveragh and Dingle peninsulas. On the sheltered mudflats behind each of these dune systems there are extensive beds of eelgrass. This marine flowering plant is a magnet for migrating brent geese, which arrive here in large flocks in the autumn and leave again in spring. In the early 1990s, I walked most of these areas at low tide, aiming to map and evaluate the area for conservation. At about the same time, the sand dunes at Inch became the centre of a controversy over an attempt to develop a golf course here. The dunes and intervening damp hollows are so extensive that it is easy to become disorientated while walking in this wild area. Legal action was threatened to prevent a golf links being laid out on the dunes at Inch following a decision by the planning appeals board, An Bord Pleanála, that the proposed development did not require planning permission. The board ruled that the proposal for the sand dunes, made famous by the film *Ryan's Daughter*, was exempt from planning control because it was started before new regulations requiring planning permission took effect in May 1994. However, an appeal by the National Parks and Wildlife Service was successful in preventing the destruction of the dunes.

In 2008, severe Atlantic storms breached the dunes at Rossbeigh, causing consternation in the local community, as a gaping hole left hundreds of metres exposed, putting low-lying homes across the bay at risk and leaving a large sandy island isolated at the end of the peninsula. The width of the breach later became more than a kilometre, which has given rise to even more concern about what the future holds in the light of rising sea levels. However, research by coastal engineers from University College Cork has found signs that the dunes are rejuvenating themselves, and the erosion may simply be part of a long natural cycle of sand movement around the bay.

Travelling by road around the head of Castlemaine Harbour near Killorglin, I headed out along the south side of the Dingle Peninsula to the village of Annascaul, where I wanted to see the home of a very famous explorer. Tom Crean was born here in 1877, but he ran away from home as a youth to join the Royal Navy for a life of adventure. He ultimately played a central role in the dramatic events of three British expeditions to the Antarctic in the Heroic Age of polar exploration. Crean was one of the last people to see Captain Robert Scott alive before his ill-fated final push to the South Pole in 1912. He also played a leading role in Ernest Shackleton's legendary *Endurance* expedition, sailing a small boat across the stormy South Atlantic before walking across the snow-covered mountains of South Georgia to fetch help for the remaining stranded seamen. After World War I, at the age of 42, Crean retired from the navy and returned to his home village on the Dingle Peninsula, where he married a local woman and adjusted to the quiet life of a rural publican.[13] I was delighted to visit the pub, named appropriately the 'South Pole Inn' and, after a pint of stout, I walked the short distance up the track to Annascaul Lough, passing Tom Crean's grave on the way. Having read extensively about his feats of endurance in the Antarctic, I was struck by how peaceful this place was, and what a contrast it must have seemed to the modest explorer after surviving for years on the ice in one of the most inhospitable places on earth.

Blasket Islands

The fishing boat carrying me to the islands departed from Dingle Harbour at first light. Rounding the iconic Slea Head into the open Atlantic, the full canvas of the Blasket Islands came into view. Dominating the foreground was the long, whale-shaped back of the Great Blasket, its grassy flanks falling away steeply to the sea. On either side, scattered in a choppy sea, were the other five main islands in this famous archipelago. Most dramatic and furthest west was the conical shape of Inishtearagh, while to the east, closest to the mainland, I could see the flat outline of Beginish. Crowded around the mother whale of Great Blasket were the three remaining islands of Inishvickillane, Inishnabro and Inishtooskert, and a number of small islets. I was travelling to the islands with a photographer working for the *National Geographic* magazine, such is the fame of these islands in the United States. This was reminiscent of a visit to the islands made in the 1930s by the well-known Dublin photographer and optician, Thomas H. Mason.

More words have been written about the Great Blasket than almost any other island in the hundreds that dot the Irish coast. This is largely because it became an almost mythical symbol of Irish native culture and language in the early 20th century. In contrast with other islands, where most traditions were handed on verbally from one generation to the next, the Blaskets produced a remarkable collection of native writers whose works have achieved the status of classic Irish literature. The approach to the main island from the sea is across the narrow sound known as *An Bealach*. Praeger related his experience here, 'in a dancing curragh on the Sound between Dunmore Head and the towering mass of the Great Blasket. The Sound is full of fish, and hundreds of gannets, from their breeding haunt on the Skelligs are at work. They dash down as thick as hailstones, and the blue water boils with their commotion.'(BS)

The broadcaster Muiris Mac Conghail described the view of the Great Blasket from the sea:

> The island looks massive enough rising out of it with a green front, and depending on the weather and light conditions a series of humps protruding from behind the green face which can alter from purple brown to very dark green. As we move closer, you become less concerned with the massive bulk of the whole landmass but rather begin to notice, to the right of the island, An Trá Bhán and the green fields with little zigzag enclosures outlining them, the first discernable evidence of habitation. When the boat moves closer to the Island you begin to notice the height of the cliffs and the fact that the Island itself is virtually raised on cliff stilts way up out of the water.[14]

On landing in the narrow creek that serves as a harbour, I walked up the village 'street', now a grassy path, between the rows of tiny stone-built houses, most now without roofs. To combat the strong winds, they were mainly built end-on to the sea with one gable end tucked into the hill. Thomas Mason commented in the 1930s that 'when one walks on the main road the chimneys are frequently on a level with one's head'.[15]

This was a time when there was still a reasonably vibrant population of about 150 people here. I could imagine turf smoke drifting from the houses, children playing in the fields, fishing nets drying outside the houses, and the island full of life and activity. But by the 1940s, due to the twin problems of isolation and emigration, most of the young people had left, and the number of islanders had dwindled to just a few hardy souls. In 1953, the final evacuation took place, when the remaining islanders moved to houses on the mainland facing the island.

Praeger visited the Great Blasket at the beginning of the 20th century, when most of the islanders were still in residence. He wrote that the village was then 'a prized sanctuary of the Irish language, a place of pilgrimage for students of the ancient tongue. Hence one is able to stay there, if fish and potatoes are deemed a sufficient diet.' His memories of the islanders are illustrated by one amusing event:

> When I botanised there one of the party was A.W. Stelfox of the National Museum who was investigating the Mollusca. The island children, consumed with curiosity, followed us about, and

> watched with astonishment the collecting of box-snails and slugs. Presently we went home to our usual dinner of one herring and potatoes. When we emerged again a deputation was waiting for us – half the children of the island, bearing cans, boxes, saucers, cloth caps and what-not all full of crawling molluscs, which they told us, a penny or two might add to our possessions. It was difficult to explain to them that only certain rarer kinds were sought for; when they realized that their labour had been in vain, the whole of their spoils was emptied at our feet, and for the rest of our stay at the cottage, inside and out, was alive with these interesting but unwelcome animals which, with misdirected energy, penetrated to every corner, and wrote their slimy autographs on wall and floor and ceiling.(WW)

When I read this passage, I guessed that the reason Praeger and Stelfox found it difficult to communicate about the molluscs was that neither of them spoke Irish (both were born in Belfast) and none of the children spoke any English. In any case, Praeger collected a large number of plants here, and in 1912 he published a note on the visit in which he described the work done by previous botanists.[16] Today, it is unlikely that any of the rarer species survive, as the Great Blasket is grazed to a short sward by flocks of hardy sheep that survive the harsh oceanic conditions by sheltering behind stone walls and buildings left by the islanders.

As I walked around the clifftops on this long island, I came upon numerous nests of fulmars, tucked into grassy ledges. First recorded breeding in 1918 on Inishnabro, with less than ten pairs, these stiff-winged birds now number over two thousand pairs on five of the main islands. Despite some visits by ornithologists in the 19th century, the birds of the Blaskets were not described in any detail until a visit in 1890 by an English ornithologist, W.H. Turle.[17] Apart from individual visits to one or other island in the group it was not until 1988 that a significant Irish team attempted a full survey of the islands' birds. Twelve intrepid ornithologists, led by Hugh Brazier and Oscar Merne and supported by the Wildlife Service and the Irish Wildbird Conservancy, set out in exceptionally calm summer weather, and a team of three landed on each of the four outer islands. Combining land-based and boat-based surveys, they managed for the first time to carry out a complete census of the seabirds on all six main islands in the group. The results confirmed that the islands held fifteen breeding seabird species, with internationally important colonies of Manx shearwaters and storm petrels, both nocturnal species that breed underground or in rocky crevices.[18]

As I walked across the main island, there were flocks of choughs above me, wheeling in the strong winds and loudly announcing their presence with distinctive 'chew-chew-chew' calls. These red-legged birds differ from other members of the crow family in that they are largely confined to the fringe of the Atlantic with Ireland as their stronghold in north-west Europe. This is because they feed principally by probing the soil and turning over animal dung in maritime grassland searching for ants, beetles, insect grubs and other invertebrates. To do this they rely on frost-free conditions along the Atlantic coastline to keep the soils soft throughout the year. Salt spray in onshore winds and extensive grazing by sheep, cattle and rabbits keep the grass short and create some bare earth, which the choughs seem to favour.[19]

I also saw a family party of choughs turning over seaweed along the driftline of the beach to collect sandhoppers that live in the rotting algae. Here, below the Blasket village, on the long white strand known as *An Trá Bhán*, the islanders played hurley and collected seaweed to transport to their fields above, as fertiliser for the potatoes that were grown around the houses. Since the final evacuation of the human population in the 1950s, the beach has become the headquarters for a huge assembly of grey seals. By 2005 the breeding population on the Blaskets was estimated at 650 to 830 animals, although over a thousand seals may haul out here in the late summer, when they are moulting. The heaving mass of enormous animals, some lying still in the sun and others sparring for space and dominance, is one of the most impressive wildlife spectacles in Ireland. Some of these seals have been tagged, with several travelling as far as the west of Scotland. One animal travelled almost 10,000 kilometres in less than a year.[20]

A century ago, the seals were an important source of food for the islanders, and the oil from their bodies kept the lights burning in every house in the long, dark winter months. They were hunted by crude, primitive methods on the remote beaches and in the caves where they bred. Tomás O'Crohan, in his famous book *The*

Islandman about the life of the community here in the 19th century, described one such hunting expedition which involved a man swimming into a cave and clubbing seals to death, whereupon they were roped up and towed out to the waiting boat.

O'Crohan also recounted a memorable occasion when he was just eight years old, as a school of 'porpoises' (in fact probably pilot whales) was driven ashore on the beach. These animals are quite tolerant of the presence of humans, and I have seen them drift around near the sea surface with the sound of their exhaling breath quite audible from the boat that I was on. O'Crohan wrote:

> When we got home, my mother cried out that the boats were coming and that some of them were making a ring around the porpoises, trying to drive them ashore. At last one of the porpoises went high and dry up on the strand. One able fellow drew his blood and when the rest of the porpoises smelt the blood, they came ashore, helter-skelter, to join the other high and dry on the sand.[21]

Boats from the island and also from Dunquin were involved in this slaughter of the whales, and when the mainland fishermen tried to take 'boatloads' of them home they were attacked by the island people until 'the men there were as bloody as the porpoises' and were driven off. O'Crohan described how the islanders then hacked up the dead animals and, when one old woman was finished, he recalled that 'she was thickly smeared with blood'.

Dingle Peninsula

The small country roads that follow the northern side of the Dingle Peninsula from Dunquin pass by the spectacular cliff coastlines of Sybil Head and Ballydavid Head, which wrap around the wide bay of Smerwick Harbour. Just to the east of the bay is the tiny Brandon Creek, which is where the famous transatlantic adventure of the Brendan Voyage began. In 1976 the maritime explorer Tim Severin led a re-enactment of the journey of St Brendan and his crew of monks in the 6th century from Ireland to North America by way of Iceland. The sailboat used was a larger version of the traditional Irish curragh, with animal hides stretched over a timber frame. Hand-crafted with traditional tools, the eleven-metre, two-masted boat was built of Irish oak and ash, lashed together with leather thongs and wrapped with a patchwork of forty-nine traditionally tanned ox hides. The seams were then sealed with wool grease.[22] I have an image in my mind of the type of frozen seas these explorers would have crossed as I read Tim Severin's book while on a flight from Iceland crossing the Denmark Strait to east Greenland.

Towering above the north coast of the Dingle Peninsula is Brandon Mountain, which is one of the highest peaks in Ireland. Praeger was in no doubt: 'Brandon is to my mind the finest mountain in Ireland'. He wrote:

> Brandon is a very wet place, and the summit is often enveloped in cloud for days at a time – in winter for weeks – yet I have always had luck there; bad mist only once in half a dozen visits, and once deterrent rain. I remember a whole week during which the summit stood up bathed in sunshine all day, while at night the black peaks rose clear against the starlit sky, looking extraordinarily near.(WW)

On the day I climbed Brandon, I was lucky too and, dropping down the steep valley on the east side of the peak, I had a wonderfully clear view of Brandon Bay and the Castlegregory Peninsula, with the Magharee Islands spread out to the north.

Castlegregory

I walked through the reedbeds that fringe the landward side of the sand dunes at Castlegregory when, suddenly, I noticed several small amphibians walking on the sandy path at my feet. These creatures are distinguished by a bright yellow stripe down the spine and the fact that they do not hop like frogs. They were natterjack toads and they were moving between the shallow ponds where they breed.

The natterjack is widely distributed along the coastal fringe of Europe, from Iberia to the Baltic, with populations in the more northern parts of its range being of more localised distribution. In Ireland, the populations are restricted to coastal sites in County Kerry on the Dingle and Iveragh peninsulas, mostly concentrated around Castlemaine Harbour and in the vicinity of the Magharees. There is a small population in south-east Wexford, introduced from

Kerry in the 1990s. The species' range is estimated to have contracted by over 50 per cent between 1900 and the 1970s.[23] As a result, the toad is classified as being 'Endangered' in Ireland.[24]Toad spawning usually begins in April, with each breeding female laying a single string of spawn. Eggs usually hatch within ten days and tadpoles develop into toadlets, usually in six to eight weeks. They favour temporary ponds where there are no predators such as fish, dragonfly larvae or diving beetles to kill the tadpoles. The species adapts quickly to new breeding sites, producing large numbers of juveniles in good years and suffering high spawn and tadpole mortality in bad years. In dry years there may be no breeding or no emergence of tadpoles, but 'boom years' can replenish a local population. The amphibians are quite vulnerable due to the small number of ponds in which they live, and drainage of these habitats would be a disaster for the species.

In 2008 a new scheme was launched by the National Parks and Wildlife Service to restore the natterjack toad to its former range in Kerry. This aimed to get farmers involved in conservation by digging ponds, all within the former range of the toad, and maintaining them, for example, through hand clearance of vegetation and by grazing the surrounding sward, in a suitable condition for toads. The programme had an ambitious target of reinstating the natterjack toad around Castlemaine Harbour and along the coastal strip west of Castlegregory on the Dingle Peninsula. Farmers entered a five-year agreement with the NPWS and received annual payments related to the number of ponds they dug and maintained.

By 2017 the pond creation scheme involved fifty local landowners, resulting in 100 new ponds created and, so far, twenty of these have shown natural colonisation by toads. To help the population expand, the NPWS has also been working with the Aquarium in Dingle and Fota Wildlife Park near Cork to develop a 'head-start' programme for tadpoles. This involves rearing tadpoles, mostly rescued from dried-up ponds, in controlled conditions and returning them as toadlets to new ponds in Kerry. This captive rearing has proved very successful, and a modest programme of translocations has also taken place.

I spoke to Ferdia Marnell, who has guided this initiative for the NPWS. He says, 'The involvement of local landowners has been key to the success of the pond creation scheme to date. Looking forward, I would like to see this programme put on a more sustainable footing as a biodiversity component of an agri-environmental scheme.' This is the rarest of only three amphibian species occurring naturally in Ireland, and we cannot afford to lose it. Expanding the number of ponds used by the toads makes the population less vulnerable in the event that one of the main ponds is damaged. It's about not putting all the toads in one basket.

North Kerry

The coast of north Kerry is often overlooked by visitors in favour of the more famous west Kerry area with Killarney at its heart. However, there are many interesting and quieter places on the coastline of Tralee Bay and points north of this. One landmark on the road to Blennerville is a restored windmill which is open to visitors. This produced flour from local grain using the constant renewable power of the wind. It was from this port that the famine ship *Jeannie Johnson* made her maiden voyage to Quebec, Canada in 1848, with 193 passengers on board. After sixteen transatlantic voyages she had transported 2,500 desperate Irish emigrants to North America. Blennerville was also where a replica of the ship was built. Over 300 shipwrights and craftsmen were involved in the construction, which took six years and was completed in 2002. The wooden ship now lies moored permanently in the River Liffey close to Dublin's Custom House, where it is open for tours.

Just north of Fenit was where the revolutionary Roger Casement and companions attempted to land guns and ammunition, at Banna Strand, in support of the 1916 Easter Rising. Casement was landed from a German submarine and promptly arrested by the authorities, but the accompanying ship, *Aud Norge*, failed to land its much-anticipated cargo of arms at Fenit, and was scuttled in Cork Harbour by its German captain to prevent the British forces taking possession of the arms cache. History repeated itself in 1984 when the *Marita Ann,* a Fenit-registered boat, attempted a similar operation on behalf of the Provisional IRA. This shipment was also discovered by the authorities and the crew arrested off the coast.

I stopped for a while to watch the wintering waterbirds in the nearby Cashen River Estuary, which enters the sea through a narrow channel just south of Ballybunion. As the tide fell, great flocks of golden plovers and lapwings circled overhead and dropped into their roosts on the expanding areas of sand that fill the estuary at low tide. Here I saw a peregrine hunting the plovers, dipping and diving as it worked the flocks to seek out a likely victim. Praeger was not impressed by what he called this 'dull country' but, nevertheless he spent 'three days along the estuary of the sluggish Cashen River in a vain search for the rare three-angled bullrush, reported to have been gathered there'. His determination to find and record the rarest of Irish plant species is well-illustrated by this quest.

> Hour after hour I concentrated my gaze on the fringe of muddy vegetation at the base of the dyke which restrains the Cashen from flooding the rushy fields on either side; and I can remember with longing my eyes turned towards the glorious outline of Slieve Mish and Brandon, towering to the southward, and running far to the west out into the ocean. But for the botanist the muddy Cashen offers an ample compensation, for it is one of only two places in Ireland where the rare dwarf spike-rush is found ... It is so rare that it is almost worth pilgriming to that dreary tract to gloat on its very insignificance.(WW)

It is clear from this passage that Praeger was uninspired by some of the places he visited, but nevertheless felt the urge to go there to complete his botanical work. I did not find the Cashen a dreary place, as I have always loved estuaries for their wide-open spaces and the opportunity to walk unimpeded across wide sand flats between the tides.

Shannon Estuary

Back on the Atlantic coast, I entered the mouth of the enormous Shannon Estuary, where the longest river in Ireland finally enters the sea. In the distance to the north was the outline of the other gatepost of this estuary at the cliff-bound Loop Head. Regarding this huge inlet of the sea, Praeger wrote:

> There is much to be seen along the fifty miles of the Shannon estuary, whether one's interests are antiquarian or biological or general, and the proper way to see it is by water, which allows of exploration on either shore. With a small motorboat and headquarters at Foynes or Tarbert or Kilrush you can have a pleasant and leisurely holiday here in a region unknown to the tourist, but possessing much quiet beauty and all

> that goes with the unspoiled Irish countryside. ... The place may become less lonely now, as these sheltered waters form the terminus of the projected Atlantic air route.(WW)

The final comment above by Praeger refers to the transatlantic crossings by seaplanes (or flying boats) which first landed at Foynes in 1937, the year that his book was published. During the late 1930s and early 1940s, land-based planes lacked sufficient flying range for Atlantic crossings. Survey flights for the flying boat operations were made by Charles Lindbergh in 1933 and a terminal was then built at what is now the Port of Foynes. As a result, this would become one of the biggest civilian airports in Europe during World War II. But its importance was short-lived and, following the construction and opening in 1942 of Shannon Airport on flat land on the north bank of the estuary, Foynes flying-boat station closed four years later. This area is now an expanding modern port linked with the port of Limerick. I spent some time surveying the shorebirds around Foynes Port to determine if they would be affected by the extension of the jetties here.

Opposite the modern port of Foynes lies a wooded island which was the last resting place of Conor O'Brien, an adventurer who was the first Irishman to circumnavigate the globe by sailing boat. Born in London in 1880 to a Limerick family, he moved to Dublin as a young man, where he practised as an architect and was a founding member of the United Arts Club, with W.B. Yeats, George Russell and George Bernard Shaw. In 1913 he climbed Mount Brandon in Kerry with George Mallory, who died a decade later close to the summit of Mount Everest. O'Brien was renowned for climbing in his bare feet with a pipe in his mouth. In 1914, having joined the Irish Volunteers, he smuggled guns into Ireland in his yacht *Kelpie*. In an apparent contradiction, he served in the Royal Navy during World War I and, on his return to Ireland, was appointed by the newly independent Irish Free State to work on fisheries development. But he was a restless soul, and in 1923 he set out from Dún Laoghaire Harbour to circumnavigate the world in his yacht *Saoirse*, returning to Ireland in 1925. He spent much of his later life on Foynes Island writing adventure novels and practical books on sailing, until he died here in 1952.[25]

The first settlers in the Shannon Estuary go right back to the Mesolithic (Middle Stone Age), when farming was unknown in Ireland. Before modern reclamation attempts, the river and its tributaries would have flowed through a network of channels, wetlands and wooded islands. In this landscape small parties of hunter-gatherers fished in the creeks, gathered wild plant food and hunted for wild pigs, hares and wildfowl in the wetlands and woodlands. From the early historic or medieval period some remarkable fishtraps survive in the mudflats. Made of post and wattle fences, these V-shaped structures were laid out across tidal channels. Fish swimming down the shore with the ebbing tide were forced by the fences to swim towards the centre of the traps, where they were caught in baskets, boxes or nets. One of the best-preserved traps still lies *in situ* on the shoreline near Bunratty on the Clare side of the estuary.[26]

Arriving at the town of Kilrush, I joined a dolphin-watching trip in a high-speed inflatable boat. The bottlenose dolphin occurs throughout the world, but the population in the Shannon Estuary is the only known resident group in Irish waters, and only one of six such groups in Europe. Each year calves are born between May and August – newborn dolphins are easily recognisable from their small size and the neonatal folds, which line their bodies. The adults vary in length but can be up to four metres long and weigh as much as 200 kilograms. They can travel at speeds of forty kilometres per hour and stay underwater for twenty minutes. We finally encountered a pod of the dolphins, bow-riding ahead of a giant tanker that had just embarked from Foynes Port. Children shrieked with delight, camera phones clicked as adults pointed avidly at anything which broke the surface. I feared that the boat would topple over as people stood in their seats and leaned over the side to get a better look at these fascinating mammals, most people seeing them live for the first time.

West Clare

The long peninsula that ends at Loop Head is the most westerly part of County Clare. I spent a few days here exploring the cliffs on the west side and the sandy bays on the south side at the mouth of the Shannon Estuary. In summer, the lighthouse on the headland is surrounded by acres of sea pinks in a rocky habitat regularly covered by salt spray from the Atlantic.

This is a wonderful location for watching passing seabirds, whales and dolphins rounding the headland from Kerry to Clare. I saw long lines of gannets commuting past from their breeding colony on Little Skellig and hundreds of Manx shearwaters that have their European headquarters in south-west Ireland. The largest coastal town here is Kilkee, which Praeger described as a recent development of the 19th century:

> Kilkee is no longer the solitude where the only sound was the voice of the seabirds in a sort of vocal accompaniment to the dirge of the 'melancholy ocean'. The waves on the coast of Clare roll and break into caverns with a sound like thunder. At Kilkee, however, under the surface of the agitated waters runs a bar which breaks the force of the ground swell. The delighted tourist can watch here the waves of the blue sea and the playful dance of emerald waters on a silver sheen of strand. Nature has furnished a walk for miles a few feet from the cliffs on a green velvet carpet all the way where this glorious accumulation of sublime grandeur can be seen.[27]

Travelling north, I came to a small harbour at Quilty, where a large bank of rotting seaweed had accumulated after a series of storms the previous winter. By now the kelp was decomposing and providing a habitat for swarms of flies and sandhoppers. Running across the heaps were dozens of small wading birds, darting here and there catching their prey. There were oystercatchers, redshanks, dunlins, turnstones and purple sandpipers. The latter species is relatively scarce in Ireland, and tends to be found mainly on exposed rocky shores rather than the sheltered estuaries that hold most of the others. A study at this harbour managed to catch and examine a total of eighty-three purple sandpipers. From the length of their bills it was thought that this group of small birds bred in arctic Canada, undertaking a marathon migration in the spring via a stopover in Iceland.[28]

Just a short distance north of Quilty is Spanish Point, which was named after the invaders who died here in 1588, when some ships of the Spanish Armada were wrecked during stormy weather. Those who escaped from their sinking ships and made it safely to land were later executed by the English authorities, who suspected that this was part of a Spanish plan to invade Ireland. Before news of the English victory reached William FitzWilliam, the Lord Deputy of Ireland, he had issued a blanket command that all Spanish found in Ireland were to be executed with their ships and treasure seized. Local legend says that the Spanish forces were buried in a mass grave, but the location of this was unknown until archaeologists, investigating the location of the wreck of *San Marcos* and *San Esteban*, announced in 2015 that they had found the grave mound at Spanish Point, where up to a thousand bodies of the Spanish sailors were buried.

Just a year earlier, the people of this coastline were reminded of the power of Atlantic storms when enormous waves battered the town of Lahinch on an eventful night in January 2014. This was the tail-end of Storm Jonas, which broke over the seafront town with such force that the waves spilt over the seawalls there. It was only the following morning that the full extent of the damage became apparent. The road had disappeared, and in its place there was a huge gaping crater just two metres from the front door of a pub. The sea walls had been wrecked and the sewerage pump house in the adjoining car park was 'blown apart'.

North Clare

Large numbers of tourists attracted to the famous Cliffs of Moher make this a busy place, often with heavy traffic. The managers of the visitor centre here recorded over 1.6 million visitors in 2019, so I avoided the area, searching for quieter refuges. It is easy to find such places, as the high cliffs extend for about ten kilometres from Fisherstreet to Cancregga Point. Reaching over 200 metres vertically above sea level, they are formed of alternate layers of shales and sandstones laid down like the pages of a book lying flat on its side. Praeger wrote, 'they are too steep to support plant life and provide few ledges whereon sea-birds can nest or rest, which gives them a savage look'. However, more detailed study later showed that the Cliffs of Moher now support the largest seabird colony on the west coast of Ireland, with over 50,000 individuals of eight species using the ledges and stacks throughout the summer.[29] This is one of the best places on the mainland of Ireland to see puffins in full breeding attire, with their large, colourful bills. Praeger wrote:

> If you want to feel small, go out in one of the canvas curraghs on a day when a ground-swell is

> coming in from the ocean, and get your boatman to row you along the base of these gigantic rock-walls. The rollers and their reflections from the cliffs produce a troubled sea on which your boat dances like a live thing, like a tiny cork, and the vast dark precipice above, vertical and in places overhanging, seems to soar up to the untroubled sky. It is a wonderful experience.(WW)

Doolin Pier is the place to catch a ferry to Inis Oírr, the easternmost of the three Aran Islands. Unlike today, a trip for Praeger to one of the many offshore islands on the west coast meant being rowed out in a curragh. There were no powerful ferries or even outboard engines. He wrote, 'I must sing the praises of the *curach* or curragh, the canvas boat used everywhere along the western coast of Ireland, for inshore fishing, lobster work and as the usual means of communication in the island regions.' Having described in some detail the construction and use of the boats, he recounted a particular incident on this coast which impressed him:

> I remember landing with George Francis Fitzgerald from an egg-shell curragh on the great

> boulder beach at Fisherstreet in Clare when the Atlantic wavers were dashing over it incessantly: our men backed the boat so skilfully that we stepped ashore without even getting splashed. (WW)

The north coast of Clare is dominated by the remarkable landscape of the Burren. This is the name given to a unique area of north Clare and south Galway that is best known for its bare limestone landscape, but it contains much more. The major habitats in the Burren are limestone pavements, orchid-rich calcareous grasslands, limestone heaths, scrub and woodlands, wet grasslands, turloughs, calcareous springs and fens. Sometimes a number of the diverse habitats of the Burren are found within the same field, where a few steps may take you from a limestone pavement, across a heath and into an orchid-rich grassland. But the significance of the Burren is the presence of so many relatively rare habitats over so large an area, offering excellent 'connectivity' in contrast with the fragmented nature of such habitats elsewhere.

I spent a few days walking in the Burren, rejoicing in the quiet places and beautiful wildflowers far from any roads or towns, listening to the sounds of nature – cuckoos calling and the wind whistling over the rocks. I began on the coast road just south of Black Head, which juts out into the Atlantic. Stretching from the road right down to the edge of the waves is the classic smooth limestone pavement with its clints and grikes with occasional large, erratic boulders perched on the surface. Here is a unique botanical mix of plants, some of which are known from the arctic and alpine regions and some from the Mediterranean area. I walked among the magenta-coloured bloody cranesbill, the deep blue of spring gentian and the delicate green fronds of the maidenhair fern, forming what Praeger called 'a veritable rock garden in spring, brilliant with blossom'.(WW)

First captivated by the Burren in the early 1970s, the artist and writer Gordon D'Arcy has spent over thirty years living on its edge and exploring its remarkable natural heritage and rich human history. In his latest book he celebrates the flora, fauna, people and places of the region with stories from his diaries and original watercolours that convey a deep affection and intimacy. One of the keys to management of the Burren is the long-established practice of transhumance – the ancient tradition of moving livestock from one grazing ground to another in a seasonal cycle. However, D'Arcy writes, 'the Burren style of transhumance is unique in that the movement reverses the norm: animals grazed on lowland pasture in the summer are moved to the uplands in the winter'.[30] To celebrate the ancient practice of moving livestock to the Burren Hills, a festival is held in October each year led by the local farming community and coordinated by the Burrenbeo Trust. It culminates in communal processions of cattle herds, farmers and many followers to the upland winterages.

The exposed limestone of the upland pasture retains the heat of the sun, 'something akin to the effects of our underfloor heating' in modern houses.[31] Frosts are rare due to the warming effects of the North Atlantic Drift, and growth continues in the winter. While the lowland areas are frequently flooded, the limestone drains readily and the hill pastures are available for grazing throughout the winter. Cattle graze on the native grasses while most of the herb species are dormant over winter, preventing them from being damaged. Grazing checks the grass, which would otherwise dominate, leaving space for more sensitive species such as orchids. Winter grazing also prevents the spread of scrub, thus keeping the famous limestone pavements open and rich in plant and animal species.

One memorable day I joined a field visit with Dr Brendan Dunford, the main driver behind the innovative Burren Farming for Conservation Programme. The current Burren Programme involves more than 300 farmers and covers over 80 per cent of the designated area of the Burren landscape. Brendan explained how this programme is different from other agri-environmental schemes. 'It has pioneered a novel "hybrid" approach to farming and conservation which sees farmers paid for both work undertaken and, most importantly, for the delivery of defined environmental objectives. Most farmers across the world are familiar with a system where they invest time and money in the rearing of livestock or growing of crops and they get paid when they sell the product at market, the price paid usually being reflective of the product quality. The Burren Programme has adopted this approach in that it pays farmers for a conservation "product" that depends on good farming practice.'

At the very northern tip of the Burren are several flat peninsulas stretching out into Galway Bay, and I walked out along one of these to the Flaggy Shore, made famous through the writing of Seamus Heaney in his poem 'Postscript': 'In September or October, when the wind and the light are working off each other, so that the ocean on one side is wild with foam and glitter ...'. Heaney is referring here to the road that divides the Flaggy Shore from Lough Murree, a brackish lagoon ' ... and inland among stones, the surface of a slate-grey lake is lit by the earthed lightening of a flock of swans, their feathers roughed and ruffling, white on white'.[32] The swans in 'Postscript' are described as having 'fully grown headstrong-looking heads tucked or cresting or busy underwater...' Gordon D'Arcy says that 'this is a vivid description of the character of mute swans to my mind'. The convoluted shoreline between the lake and the sea has colourful clumps of thrift, its bright pink flowers mixed with white blossoms of sea campion and a backdrop of the yellow-flowered sea radish. I didn't see any of Heaney's swans, but I did hear the distinctive repetitive whistle of the whimbrel. Migrating north from their winter quarters in West Africa, these long-billed waders often stop off here in May on their way to their breeding grounds in Iceland.

Marine biologist Dave McGrath knows this area well, and has a special interest in the Carrickaddy reef. At low spring tide he finds an enormous number and variety of marine plants and animals here, probably due to the combination of sheltered conditions and the multiple niches available in the rockpools among heavily eroded limestone. Here he has found porcelain crabs, iron crabs, shore crabs, velvet crabs, squat lobsters and purple sea urchins, all kinds of molluscs, rock-pool fish including pipefish, scorpion fish, blennies and gobies. There are also some rare seaweeds here too.

Aran Islands

My first visit to the Aran Islands was very memorable. I was the passenger in a small, single-engined aircraft piloted by Con O'Rourke, who later wrote a *Nature Guide to the Aran Islands*.[33] We flew in dense mist, using only the instruments, out over the Atlantic but, as we descended below the gloom, a remarkable landscape emerged below us, looking like a network of lace. The three limestone islands are covered by thousands of small fields, divided by stone walls, with winding roads and paths between them. Botanist Cilian Roden wrote that navigation on the ground, 'especially on the two smaller islands, bears comparison to making your way about the blocks and alleys of an unfamiliar city'.[34]

Praeger had a special affection for the Aran Islands that form a line across the outer part of Galway Bay. He gave an excellent introduction:

> If I wished to show anyone the best thing in Ireland I would take him to Aran. Those grey ledges of limestone, rain-beaten and storm-swept, are different from anything else. The strangeness of the scene, the charm of the people (I don't refer to the rabble that meets the steamer), the beauty of sea and sky, the wealth of both pagan and Christian antiquities, the remarkable vegetation (without a parallel in western Europe save in the adjoining Burren of Clare) – all these help to make a sojourn in Aran a thing never to be forgotten.(BS)

Despite this appreciation, Praeger gave little attention in *The Way that I Went* to the unique flora of the islands, relegating discussion of it to his section on the similar flora of the Burren, although the islands are scarcely mentioned there. Several plant species that were probably common in Praeger's time are now rare in the Aran Islands and in the whole of Ireland. However, in 1987, botanist Tom Curtis was surprised to find several rare and threatened arable weeds, such as cornflower and darnel, growing from the thatched roofs of houses on the islands, where their seeds had germinated from the rye straw traditionally used for thatching.[35]

I walked across all three of the islands, marvelling at the unique landscape with cliffs dropping vertically to the Atlantic Ocean and a rich flora that grows on the limestone grassland. The Aran Islands are administratively within County Galway but resemble more closely the rocky landscape of north Clare, with a marked oceanic influence. The three inhabited islands – Inis Mór, Inis Meáin and Inis Oírr – cover a total of 4,330 hectares and have supported farming communities for over 4,000 years. This has left behind a rich cultural legacy most dramatically seen in the spectacular great forts on the islands and the dense web of field wall systems.

Today, the agricultural system on the islands is mainly low-intensity production of cattle and sheep, with a small area of tillage for on-farm use. The soils are thin with low fertility, and livestock are frequently exposed to high winds. The farms are tiny, with over 85per cent being smaller than 20 hectares in area, less than half the national average size. Farms are highly fragmented, often made up of several different habitat types, such as calcareous grassland, limestone pavement and machair, a rare type of coastal grassland. The cattle are grazed in summer on the richer *fear glas* (green grass) on the more sheltered northern side of the islands. The mild climate and dry fields mean that cattle are not housed but wintered on specific areas of the farm. This is critical in ensuring the presence of such a distinctive flora as, without adequate winter grazing, bramble and blackthorn would become dominant. Cattle diets are often supplemented with hay, which is still cut by hand and often dried on the surrounding walls.

While some past agri-environment schemes benefitted the islands through the provision of economic support to farmers, they failed to adequately address a number of conservation issues, including the maintenance of priority habitats. Farmers on the islands felt a different approach was required. The AranLIFE project got going in 2014 and worked with farmers across the three Aran Islands to improve the conservation status of habitats of international importance. This scheme was prompted by the success of the Farming for Conservation Programme in the Burren. A key measure here was improved access to the many scattered land parcels to facilitate management, including grazing. Scrub and bracken were controlled while livestock management was enhanced by installing new water features, and hundreds of existing rainwater tanks were repaired. This secures the harvesting of rainwater, which would otherwise drain away through the limestone rocks. The AranLIFE project ended in 2019, but this work is continuing under a new project called Caomhnú Árann, run by the same team.

On the largest island of Inis Mór, I walked along the dramatic cliffs on the west side near the stone fort of Dún Aonghasa. Thousands of seabirds, mainly kittiwakes, guillemots and razorbills, lined the ledges formed by softer shale bands between the limestone strata far below. In previous centuries these birds and their eggs were harvested by the islanders to supplement the food produced on their small farms. In his classic book, *Stones of Aran: Pilgrimage*, Tim Robinson described how this hunt was conducted:

The men would walk across to the cliffs at dusk with the rope, which was often a communal investment. One end would be tied around the cliffman's waist and between his legs, and the other made fast to an iron bar driven into a crevice or wedged to a cairn on the clifftop. A team of up to eight would lower the cliffman, guided by signals from a man stationed out on a headland from which he could watch the progress of the descent. The cliffman would carry a stick to keep himself clear of the cliff face while swinging on a rope and wedges to help him round awkward corners of his climbs. Having reached the chosen ledge the cliffman would remain crouched in it until darkness came. When all the birds had flown in from the sea and settled down to roost he would begin to crawl along, and would silently murder the first bird he came to ... He would then move on, pushing the dead bird before him until it was up against the next victim, which thus would not feel his hands until it was too late. The dead birds would be strung on a cord by running a loop around the neck. At dawn the cliffman

> would be hauled up again, bent and rigid with cold and cramp.[36]

Today the people of the Aran Islands do not have to subsist over winter on the dried flesh of seabirds, but they have managed to protect a unique landscape with much of its rich biodiversity still intact. As I walked along the cliffs, I marvelled at the colourful wildflowers, thriving bird populations and the power of the Atlantic Ocean far below.

Galway Bay

I visited Kinvara, on the south side of Galway Bay, for the festival of *Cruinniú na mBád* (Gathering of the Boats). This annual event celebrates the living tradition of the sailing Galway Hookers and related boats. Up to a hundred traditional boats, once the workhorses of the coast of Connemara and north Clare, gather here for a weekend of racing and celebration. There are few more inspiring sights at sea than a fleet of Galway Hookers in full sail with the landscape of Connemara behind. They developed as cargo boats ferrying turf and other local goods across Galway Bay to the Aran Islands, to Galway city and to north Clare, to avoid the long road haul by horse and cart. None of the latter regions had their own turf bogs. To make the fastest delivery, the boats would race each other. By the 1960s, improvements in road transport and the widespread use of bottled gas as a fuel on the islands led to the demise of the working Hookers, and many were left to rot on beaches and quaysides.

In the late 1970s, a few dedicated traditional boat enthusiasts stimulated a revival of the boat-building tradition in Connemara, rebuilding some of the original boats and constructing others from scratch. This led to the sport of Hooker racing, with a series of annual regattas. Today, the Galway Hookers Association organises races all over Connemara and the Aran Islands. With its trademark red sails and wide-beamed black hull, the Hooker is now a well-known symbol of Connemara. Since its revival in 1979, *Cruinniú na mBád* has become one of the most unique and successful weekend festivals in the west of Ireland. The event also promotes the art of traditional music and the culture that surrounds it, including Irish dancing, *sean nós* singing, the Irish language and more.

During the festival I had arranged to meet my friend Cian de Buitléar, who is skipper of *Star of the*

Robert Lloyd Praeger (1865-1953) explored almost every part of the Irish coast in his quest to record its natural history. His wife, Hedwig, accompanied him on many of his adventures. (*courtesy of the Royal Irish Academy*)

Harbour seals are found in many sheltered bays and inlets around the Irish coast. The pups are born in mid-summer and can swim from birth. (*courtesy of John Fox*)

The compass jellyfish is one of the larger floating animals in Irish inshore waters. It is named because of the markings which resemble the cardinal points on a compass. (*courtesy of John Fox*)

Rockabill is located off the north Dublin coast. The rocks around the lighthouse hold a huge colony of nesting terns, small migratory seabirds that breed here in summer.

Kittiwakes are the smallest of the gulls, often gluing their nests to narrow ledges or vertical cliff faces. They have declined rapidly due to food shortage caused by overfishing. (*courtesy of Brian Burke*)

Brent geese have adapted to live within urban areas where they feed, not only on estuarine plants but also on sports pitches and other amenity grasslands.

Yellow horned-poppy at the Murrough in County Wicklow. This is a typical shingle beach plant.

The ruins of Black Castle in Wicklow Town silhouetted against a dawn sky. The castle was repeatedly burnt and destroyed in the wars between invaders and residents.

Puffins are probably the best-known seabirds with their colourful bills and clown-like appearance. They nest in underground burrows, mainly on offshore islands. (*courtesy of John Fox*)

Tourist boats below the cliffs of Wexford's Great Saltee Island ply the waters once used by smugglers and pirates.

A storm lashes the seafront of Dunmore East in County Waterford. (© *Liam Blake*)

The impressive battlements of Charles Fort guard the entrance to Kinsale Harbour in County Cork. This was used as a garrison from the seventeenth century up to the Civil War of the 1920s.

Common dolphins off the coast of West Cork. These marine mammals often move about in large schools and are frequently attracted close to boats. (*courtesy of Tatiana Lie Kumagia*)

The remarkable monastic remains on Skellig Michael seen from the air. This remote spot was occupied by a group of monks from the 6th Century AD. (© *Liam Blake*)

The grey seal colony below the old village on the Great Blasket is one of the largest in Ireland. The seals were once hunted by the islanders but have multiplied with legal protection. (© *Liam Blake*)

A sea arch at the Bridges of Ross in County Clare shows the power of the Atlantic and the pressure that has caused these folds in the rock over geological time.

A cottage and stone pier in Connemara. Hundreds of small piers were built in the late nineteenth century by the Congested Districts Board to stimulate fishing on the west coast.

A tree in Connemara shows the effects of almost constant winds on the Atlantic coast.

Keem Bay in Achill Island was the scene of historical mass killing of basking sharks that almost caused their extinction.

Achill Yawls are wooden sailing boats with an old-fashioned gaff rig. In recent years there has been a revival of interest in this traditional vessel with a workshop at Mulranny to build new yawls.

The remains of a century-old whaling station on South Inishkea Island with the deserted village in the background.

Grey seal with a young pup. The pups are born in the autumn on remote beaches and in caves where they are far from disturbance. (*courtesy of John Fox*)

A walker on the beach at Enniscrone, County Sligo watches surfers in the waves. The north-west of Ireland offers some of the best surfing conditions in Europe.

Oystercatchers are among the most easily recognised waders on the Irish coast. They feed on a variety of shellfish and worms, in estuaries, rocky shores and coastal grasslands. (*courtesy of John Fox*)

Tory Island, lying 12 kilometres off the Donegal coast, is one of the more remote inhabited offshore islands. It is one of the last refuges in Ireland of the threatened corncrake. (© *Liam Blake*)

Malin Head is the most northerly point of the Irish mainland and windiest site in the country. The steep slopes in the fields represent ancient shorelines when sea level was much higher than today.

Dunluce Castle County, Antrim is perched on the edge of basalt rocks and can be reached by a bridge from the mainland. (© *Liam Blake*)

Flooding at Clontarf in Dublin City. Sea level rise and an increase in frequency of storms is causing flooding in many low-lying parts of the coast. (*courtesy of John Fox*)

The lighthouse on Hook Head, County Wexford, is thought to be the oldest in Ireland or Britain, dating originally from the 5th century. The rocks around the shore are full of large fossils, and the seabed is rich in marine life.

The author sailing past the Fastnet Rock, the most southerly point of the Irish coast.

West, built in 1994 and named after an older boat that foundered off the Codling Bank on the east coast in 1948.[37] Cian gave me the helm for a few minutes but I scarcely had the strength to turn the rudder. A modern racing yacht can change direction in a short distance, but the massive Hooker takes a fair bit of effort to tack. There were races for the four types of Galway Hookers, including the *Bád Mór* (big boat), which is the largest vessel, measuring over 12 metres in length, the *Leath Bhád* (half boat), the *Gleoiteog* and the diminutive *Púcán*. The smaller boats were originally used in shallow water to gather seaweed for fertiliser or production of kelp for export. As we arrived back at the quay, a fiddler and tin whistle player tuned up on the deck of Cian's boat. During the festival, all the pubs in Kinvara were alive with the sounds of music, as this event attracts the very best traditional Irish musicians.

Travelling around inner Galway Bay, I passed by the extensive saltmarsh of Tawin Island, well grazed by sheep but still rich in plant species such as sea pink, scurvy grass, sea lavender and yellow horned-poppy. I also found the beautiful herb sea wormwood, a greyish plant that flowers in late summer. It has a wonderful aromatic perfume and is quite localised but abundant around Galway Bay. None of the Galway Bay saltmarshes are associated with sand dunes, and this may have arisen because of marine flooding of glacial deposits rather than by vegetation colonising mudflats, as happens on the east coast. There are places at sea level where the saltmarsh soils lie on top of peat, similar to that which occurs on the lowland blanket bogs of Connemara. This shows that there was a gradual submergence of the land as sea levels rose after the melting of glaciers at the end of the last Ice Age. Tawin is also a great place to watch the flocks of wintering wildfowl and waders like curlew and lapwing as they gather here at high tide. Up to fifty harbour seals haul out on the storm beach, where they are rarely disturbed.

On the north side of the bay is Galway, the ancient city of the tribes, strategically located where the River Corrib empties a vast quantity of water each day into Galway Bay. This marks the junction between limestone rocks to the east and south and the granites of Connemara to the west. I went there in 2009 to see some of the fastest yachts in the world, as an estimated ten thousand people gathered to welcome the Volvo Ocean Race fleet into Galway Bay. In the early hours of the morning bonfires were lit on the Aran Islands as the yachts finished this leg of a race that circles the globe. I joined the crew of a Galway vessel as a spectator to watch these high-speed racing machines arriving into the bay. With enormous black sails, they sped past us towards the finish line, setting up a wash that shook our own boat.

Just off the Salthill shore of the city is Mutton Island, connected to the land by a narrow neck of slippery rocks at low tide. I began a survey of shorebirds here some years ago by rowing out from Nimmo's Pier to land on a slipway below the long-abandoned lighthouse. The door to the tower lay open, so I was able to climb the winding stairs to the top and poke my telescope out through the broken windows. Below me at high tide the rocks around the island held a large wader roost, with up to 2,000 birds, including oystercatchers, curlews, dunlins and ringed plovers, just one or two kilometres from the city centre. As there is little human disturbance here the birds continue to use the island, where a large sewage treatment works for the city has been built.[38]

South Connemara

Praeger never drove a car in his life but he was fortunate to have the use of the most extensive network of small railways ever built in this country. One such line crossed the boggy region of south Connemara, and he revelled in the beauty seen from a train journey here in 1896.

> One of the most important Irish railway enterprises of recent years was unveiled last Spring by the opening of the new railway from Galway to Clifden. Leaving the capital of the West, the line skirts the wooded shores of Lough Corrib as far as the little town of Oughterard, and then strikes westward across a picturesque wilderness of lake and bog to meet the Western Ocean. ... Many rare plants were obtained both in sea and on land, many a strange insect that will delight the entomologist, many a pleasant impression of the cheery peasants, clad in their rough homespun, with hearts as warm and bright as the April sun that streamed down day after day, who told us wonderful folk takes and sang us crooning songs, in their own Irish tongue.[39]

Praeger was also a strong advocate of walking as much as possible so that he could study the vegetation and record interesting plants. A century later, I too was walking the coast of Connemara, the large region that lies west of Galway city and Lough Corrib, occasionally taking a dip on the numerous beaches. I camped overnight on a remote, sandy peninsula and looked west over the ocean until the round, orange sun dropped below the horizon. Tim Robinson described this fringe of Connemara as 'a rope of closely interwoven strands flung down in twists and coils across an otherwise bare surface'.[40] Here, much of the shore is covered in a thick blanket of brown seaweeds, obscuring the rocks at low tide. Snorkelling among the weed at high tide is like exploring a forest on land. Above my head the canopy of seaweed swayed back and forth in the tidal currents and sunlight poured through the gaps, illuminating the sandy bottom.

In between the rocky outcrops there are many sandy shores and small coves. Some spectacular beaches surround the village of Roundstone, which gives a fine view of the Twelve Bens, a mountain range which forms the boundary of South Connemara. Praeger admitted that he 'dwelt long on Roundstone, because it is at that spot a kind of concentration of interests, as well as a delightful combination of hill and moor, sea and lake, rock and sand':

> Dog's Bay is one of the greatest attractions that Roundstone offers. A narrow mile-long granite island set half a mile offshore has got joined to the mainland by a sand-spit, which forms two curving bays – Dog's Bay and Gorteen Bay – set back-to-back and filled with the clearest of Atlantic water. The sand, excessively white, is in itself remarkable, for it is formed not of quartz grains, as is usual, but of shells – mostly the tiny perfect tests of Foraminifera (of which no less than 124 species and varieties have been found here), the rest being the comminuted shells of the more familiar Mollusca.(WW)

After a long walk above rocky shores I reached Bunowen Pier on the south side of the Slyne Head Peninsula. It was a welcome surprise to find the Connemara Smokehouse here, run by the Roberts family. More than forty years ago, John Roberts, a lifelong fisherman, purchased a 1946 fish-smoking kiln

– fondly dubbed 'Old Smoky'. The unique flavours created by the kiln helped the smokehouse become famous for producing fine smoked fish. Since the 1990s, when John's son Graham and his wife Saoirse took over the reins, the family smokehouse team has grown to include their four children.

I bought some smoked mackerel and a delicious side of smoked albacore tuna caught on pole and line (troll-fished) in Irish waters and smoked here. The smoking process has been perfected over generations of the family business. Beechwood smoke is fanned through the fish for up to ten hours, depending on the size. A similar amount of time is required to dry the fish. Because of the age and simplicity of the kiln, even the weather conditions from day to day affect the way the fish are cured. Having undergone some major surgery and now enjoying some tender loving care, 'Old Smoky' still has pride of place at the Connemara Smokehouse.

Emerging from the smokehouse I noticed that a small inshore fishing boat had just tied up at the end of Bunowen Pier. My curiosity aroused, I wandered down to find out what they had caught. Skipper Pat Conneely was already unloading boxes of live lobsters and crabs that he had captured in lines of pots some ten kilometres off the coast. Pat told me that the pots were laid near the Skerd Rocks, a cluster of jagged reefs which mark the change from the shallow coastal ledge running out from land to the steep underwater cliff that slopes rapidly down to depths of over a hundred metres. This underwater landscape is a rich fishing ground for lobsters, just as it was a century ago, when the local fisherman-naturalist Séamus Mac an Iomaire sailed out with his father from the island of Maínis to catch fish here. In his unique book, *Cladaigh Chonamara (The Shores of Connemara),* Mac an Iomaire later accurately described the ecology of a wide range of marine creatures such as *an gliomach* (the lobster):

> Many are caught in pots in summer and autumn. During that time they come in large shoals in from the strange deep. They are big and small depending on age. You would often see the young ones swimming on top of the water during fine weather. They are only as big as a shrimp, about two inches long. It's hard to notice them as they are the same colour as the water; nevertheless,

other fish slaughter them. Only a small number escape. Indeed, if it happened that all of the seed matured, then the sea would be full of lobsters. Nature prevents this. It's wonderful the way that it can keep the fish under control so that they won't be too plentiful and won't be destroying each other.[41] This remarkable understanding of the key principles of population dynamics was by a man with little formal education but with a deep knowledge of the natural world, learned over years of close contact with the sea and its creatures. Back on Bunowen Pier, Pat gave me a large bag of crab claws for my dinner but, with the typical generosity of his people, he would not accept any payment for these delicacies.

Mannin Bay

Following the low rocky shores towards Slyne Head, which marks the most westerly point of Connemara, I found secluded sandy beaches hidden away from the average tourist. I scrambled from the beach at Mannin Bay up a steep sand cliff and onto a level plain of sandy grassland that stretches away along the coast. This is a unique type of sand dune known as machair that occurs at a few places along the north-west coast of Ireland from Galway Bay to north Donegal. Unlike normal sand dunes, the machair is almost flat, sculpted by the constant winds from the Atlantic Ocean. In fact, the word machair comes from the Gaelic *maghera*, meaning a plain. During the winter, ocean storms carry great clouds of sand from the beach, spreading this across the machair. The sand is rich in calcium from the broken seashells that litter the shore. The grasslands are almost universally grazed by cattle, sheep or horses. The rich mixture of plant species makes for a mineral-rich diet, and the animals are healthy and hardy as a result.

Back in the 1980s, machair attracted the attention of a group of naturalists in Ireland. Tom Curtis led a team of botanists mapping the boundaries, flora and condition of over a hundred such sites along the west coast.[42] A typical suite of lime-loving grassland plants, including red fescue, plantain, daisy, bird's-foot trefoil, lady's bedstraw and white clover, is found in virtually all machair sites. Occasionally, a flush of beautiful pyramidal orchids provides a highlight in the grass. Walking over the flower-rich grassland, I dodged the cattle that graze this highly nutritious sward. Suddenly a loud 'chew-chew' call announced the arrival of a family of choughs. They whirled and dived in aerobatic manoeuvres that would qualify them for any airshow.

Similar habitats in the Outer Hebrides of Scotland were found to have some unique populations of breeding waders. So, forty years ago, with several other ornithologists, I decided to have a look at the Irish machair sites identified by the botanists to see if they were similar. Over a period of months in the summer we walked over fifty sites in six counties. Here we found lapwings, circling about the sandy grasslands with their loud mewing calls, snipe drumming above the marshy areas, tiny dunlin with black plumage on their bellies and ringed plovers nesting among the shells on the strandline. Occasionally, we found a pair of oystercatchers, our attention drawn by the loud piping noise that they made to distract us from their chicks. The richest sites were those with both wet and dry machair in close proximity.

Today there have been dramatic changes in the breeding wader populations of the machair, and the sandy grasslands of Mannin Bay are no exception. Dunlin have all but disappeared, and lapwing declined by about a third in the following twenty years.[43] Although lapwing have attempted a comeback, with thirty-two pairs found here in 2019, few young survive, and productivity remains relatively low. Today, the overall numbers of breeding waders on these sites have declined by a further 80 per cent, and many sites are virtually deserted. The machair itself is under a number of serious threats, including pressures for recreation, subdivision by fences facilitating more intensive grazing and the application of artificial fertilisers. Changes in the grassland structure are evident, with much lower coverage of tussocks, which many waders require for nesting. This is thought to be the result of increased grazing pressure in recent decades, with sheep densities markedly higher at numerous sites. These changes in habitat structure are also likely to exacerbate predation pressures, where scavengers like foxes and hooded crows can more easily find the nests. These threats can change the nature of flora and the birdlife permanently.[44]

Walking the beaches in Mannin Bay, it is clear that this is a most unusual sort of 'sand'. Looking closely at a handful of the white material, I could see that it is coarse, branched and coral-like in appearance, and it felt rough under my bare feet. In fact, it is not sand at all but a material called maërl that is created by several seaweed species which concentrate the minerals from seawater with calcium forming tubes around the living plant. When alive they are bright pink in colour, but when broken up by storms and washed ashore they become bleached and white-looking, like tiny interlocking bones. So much of this material is present that it makes up the entire beach in some places, like the 'coral strands' of Mannin Bay.

Once used by west coast communities as a fertiliser on lime-poor soils, maërl is mainly composed of calcium, but it contains much higher proportions of magnesium, iron and other elements than ground limestone. It also has a more porous structure than the normal lime fertiliser, which may help bacteria to break it down more easily in the soil. It is found in some sixty offshore banks on the bottom of shallow inlets along the west coast, from Roaringwater Bay in West Cork to Mulroy Bay in north Donegal. In Europe as a whole, these algae are found from the Mediterranean to the Arctic Ocean, but are best known from Brittany in north-west France, where the word *maërl* originated. Maërl creates a structurally complex habitat on the seabed that is colonised by a wide range of other plants and animals, especially algae (seaweeds) and crustaceans (crabs, shrimps and amphipods). So it is of high conservation value, and when there are proposals for large-scale commercial exploitation of the maërl for its value as an agricultural fertiliser, I feel that this is the coastal equivalent of mining the bogs and sending the turf up in smoke. Short-term use creating long-term damage.

As I walked the beaches of west Connemara I picked my way between large piles of partly rotted seaweed cast up on the strandline by recent storms. But there were also long blades of what appeared like the grass from a meadow. In fact this is a flowering plant called eelgrass that grows in shallow water and is frequently torn from its roots by stormy seas. When seen growing beneath the sea these seagrass beds look exactly like an unmowed meadow, with the leaves waving back and forth in the current. Like all green plants they depend on sunlight to photosynthesise and grow, so clear water and shallow seas provide exactly the right conditions. Just like a hay meadow on land, healthy seagrass beds support a wide diversity of animal life, both above and below the seafloor. They offer safe nursery areas for fish like cod and herring and shellfish such as scallops. Like leopards stalking through the African grasslands, sharks and rays also use the cover of these marine meadows to hide from their prey. Globally, at least fifty species of fish live in or visit seagrass beds, and about one-fifth of the world's biggest fisheries are supported by seagrass meadows as fish nurseries. Seagrass habitats support thirty times more animals, mainly invertebrates, than other nearby habitats. Importantly, in the current climate crisis, seagrass meadows store carbon as effectively as forests, at about 400 kilograms of carbon dioxide per hectare each year.

Seagrass communities are declining across the world's oceans, and Ireland is no exception to this. Globally, estimates suggest that we are losing an area of seagrass around the size of a football pitch every 30 minutes. One of the main causes of this decline is pollution draining into the sea via rivers from agriculture, forestry and urban settlements on land. Nutrient enrichment through the discharge of phosphates and nitrates causes other more vigorous plants to grow on top of the seagrasses, blanketing them from the vital sunlight. Declines of this sort have been noted in bays all along the west coast, from Castlemaine Harbour in Kerry to Mulroy Bay in Donegal. In Wales, a new project called 'Seagrass Ocean Rescue' has been launched to trial methods of restoration. Seeds are collected from the eelgrasses when they mature, and these are grown in laboratory conditions so that the plants can be later sown back out on the ocean floor.

Inishbofin

Following the indented coastline of northern Connemara, I passed through the lively tourist town of Clifden and came, after a short distance, to the extensive strands opposite Omey Island and the coastal village of Cleggan. This was the location of a terrible tragedy in 1927, when twenty-six fishermen, including some from the island of Inishbofin, were caught out in their curraghs in a sudden storm and drowned. Known as the 'Cleggan Disaster', this same storm also took a toll on fishing communities in several other west coast locations.

A regular ferry runs from the pier in Cleggan to Inishbofin, entering one of the most sheltered natural harbours on any of the islands. The entrance is guarded by the impressive ruins of Bosco's Castle, so-called because it was once occupied by the Spanish pirate, Don Bosco. From the pier where the ferry lands, I walked around to the castle to explore the remaining walls and substantial defences. According to local lore, Bosco had an alliance with the O'Malley clan, which ruled much of this western seaboard in the 16th century. In the mid-17th century, following the Tudor conquest of Ireland, the castle was garrisoned by Cromwell's soldiers when Inishbofin served as a transportation centre for priests.

The origins of the island are rooted in a legend which is often quoted to explain its name *Inish Bó Finne*. The most common version relates how two fishermen, lost in fog, landed on the island and lit a fire. The flames broke a spell and the mist lifted to reveal an old woman driving a white cow along a beach, which ran between a lake and the sea. She was seen to strike the cow, whereupon it turned to stone. Another story tells that the old woman and the cow emerge from the lake every seven years to forewarn of some impending disaster. The lake in question is *Loch Bó Finne* (Lake of the White Cow) in West Quarter village.

I have a soft spot for Inishbofin, as my father found a safe anchorage for his yacht in this harbour during a major storm when he undertook a circumnavigation of Ireland in the 1990s. On my own visit, as I stepped off the ferry, I could hear the distinctive 'crake-crake' call of a corncrake from the fields behind the houses. Now largely confined to some remote islands and coastal fields, this threatened bird is clinging onto Ireland by its wingtips. It makes an impressive migration each year to the tropics south of the Sahara Desert, returning in May to search for long vegetation such as nettle beds, where it will hide before nesting in the growing meadows. Early summer mowing of hay or silage often leads to the death of the flightless chicks and failure of second breeding attempts too.

South Mayo

Killary Harbour is a long, narrow inlet, extending sixteen kilometres in from the Atlantic to its head at Leenaun, below the famous Aasleagh falls. It forms the border between Galway and Mayo and boasts some of the most spectacular scenery in the west of Ireland. Praeger described it as 'the best example in Ireland of a true fiord – a submerged river-valley which has been greatly deepened by ice-action ... but, as in the manner of fiords, its entrance is less deeply excavated than its middle part, and requires careful navigation.'(WW) The main part of the inlet is extremely deep, reaching down to 45 metres at its centre. This offers a very safe, sheltered anchorage among high mountains to the south and north. It is a centre for shellfish farming, with lines of floats and hanging ropes that are used to grow mussels. Clams and mussels grown in Killary Harbour are sold at the Westport Country Market every Thursday morning.

Entering County Mayo, Praeger was also familiar with the coastal strip north of Killary Harbour; 'a flat sandy stretch with shallow lakelets and a curiously mixed flora intervenes between the Atlantic and the steep gable of Mweelrea – a lonely and fascinating place, known as Dooaghtry'.(WW) This description hardly does justice to a wonderful natural mix of beaches, dunes, saltmarsh, machair, lakes, fens, native woodland and rocky outcrops that still retains much of the magic and fascination hinted at by Praeger. I have stayed here many times as a guest of my friend David Cabot, who lives in the centre of this amazing landscape. He has shown me some of the hidden wonders of this place. The lakes, which are surrounded by species-rich alkaline fen and sandy grassland, have a rich community of aquatic plants including few-flowered spike-rush, carnation sedge, water mint, grass-of-Parnassus and the marsh helleborine orchid. The lakes are also used by migratory whooper swans that fly in from Iceland in the autumn, their loud trumpeting calls echoing across the nearby machair. On an early morning walk I saw an otter emerge from one of these lakes and, with its characteristic weasel-like gait, it ran across the rocky outcrops and disappeared into the reeds in a neighbouring waterbody. On the soft sand of the beach I picked up its tracks, looping up and down the shore as it searched for stranded prey on the driftline. In the rest of Ireland otters are mostly nocturnal, but here in the west of Ireland, where they are little disturbed by the local residents, they emerge to feed during the day. Usually associated with freshwater, these wonderful mammals are equally at home in the sea, where they swim among the seaweed and frequently bask on the shoreline.

Behind the wind-battered dunes is a classic area of flat machair plain, grazed over centuries and formerly a top breeding site for waders like lapwing and dunlin until the national population crashed in the late 20th century. One of the factors which has protected this complex area from development or tourism is that to reach it you have to wade or drive through the shallow tidal outlet from the lakes, and few people are willing to do this. When I first visited this unique area in the 1970s there was a large pyramid of sand on the beach. This was a famine graveyard, where countless victims had been buried in the mid-19th century. Climbing up the sides of this iconic structure I could see bleached human bones being eroded from the sand and I thought of the great hardship faced by the poor people of this area a century and a half ago.

Close to the road to Louisburgh I walked down to see a historic clapper bridge that was built in 1863. This is a low, curving construction, some fifty metres long and formed of thirty-eight boulders spanned by flat slabs of stone. I was able to walk on the bridge beside a wide but shallow ford across the river. The former village here was known locally as 'The Colony', but its past is seldom talked about, and few are willing to recall the painful memories of its brief existence. During the famines of the 1840s many local people converted from Catholicism to Protestantism in return for food to keep them alive. In 1853, the landlord Lord Sligo granted the local Protestant minister an acre of land here, on which to build a new church. As this was in an extremely remote area and there were no roads to it, numbers began to fall and the congregation eventually petered out. The Society for the Protection of Rights of Conscience took control of the church and built ten or twelve houses for the converts together with a school for its inhabitants. However, the local people strongly objected to the presence of the society and after only twenty years it disappeared from the district and the church was demolished in 1927. All that remains of the Colony today are the ruins of three houses.

Inishturk

From Roonagh Quay, I caught a local ferry to the island of Inishturk to find some space and rest. Slightly less well-known than either of its neighbouring islands, Inishturk is sandwiched between Inishbofin to the south and Clare Island to the north. Praeger and his wife, Hedwig, spent 'an interesting week' in the island in 1906. His comments on the accommodation illustrate how his new wife was unshakeable in her support of the naturalist and his work:

> There is no inn there but, by the kindness of F.G.T. Gahan we secured the use of a shed belonging to the Congested Districts Board, perched on a low rock with deep water half surrounding it. In bad weather the beating of the rain on the galvanised iron roof, combined with the roar of the waves just outside the door, made a wonderful noise at night. When we arrived, after a lively passage from Renvyle, I was sent off to botanise for at least two hours, and when I returned my wife had removed from the floor a half-inch compacted layer of Portland cement, herring scales, petroleum and sawdust. Then we settled down, surrounded by coils of wire, boxes of dynamite, and bags of cement, and we fried fish, baked bread and cooked bacon and eggs on a small pan over a smoky stove.(WW)

My own accommodation in a comfortable guesthouse was much more amenable. The hosts were very welcoming and even took me with them to a musical evening in the community centre. Today, the island has a permanent population of about sixty, although in 1861 more than twice as many people lived here, and they were then entirely Irish-speaking. I wandered right around the cliff-bound island, regularly leaving the circular road, following choughs and puzzling over heaps of rocks that might have been the remains of deserted cottages. The western side often experiences huge Atlantic waves, and there is a blowhole near the north-east tip of the island. The waves enter a cave below sea level and the pressure forces a fountain of seawater up through the boulders on top. I was greatly impressed by the Gaelic football pitch wedged between steep rocks in the centre of the island. This must be the most remote sports pitch in the country, and is well sheltered from the regular strong winds.

I got used to the constantly patrolling fulmars and gulls along the cliffs but I was surprised when a large dark bird started to hover over me and made a few attempts to chase me away from one of the more remote parts of the clifftop. This was a great skua, the newest addition to Ireland's impressive list of breeding seabirds. It was traditionally confined to the northern islands of Scotland, but has recently been expanding its range as it adapts to preying on fish discarded by vessels at sea. These large, aggressive birds will chase other seabirds, forcing them to drop the fish they are carrying, which they will then steal. Later in the season the skuas will patrol the cliffs searching for seabird chicks to prey on. In 2010 breeding by skuas was reported on Rathlin Island in Northern Ireland, while more of these birds were seen prospecting cliffs in Donegal. The latest surveys for the National Parks and Wildlife Service estimated breeding by thirteen to fifteen pairs in the Republic in 2015–18, but Stephen Newton of BirdWatch Ireland thinks that there may be a minimum of about twenty-five pairs of skuas by now on the island of Ireland. They have now been confirmed breeding on a number of other islands around the coast between Galway and Antrim.

Praeger repeatedly mentions the many values of the curragh, the traditional boat design in remote western islands. 'The curragh serves as a tent also, for the lobster-men often spend a night on homeless islands, sleeping comfortably under the overturned boat whatever the weather may be'. He recalled an event where one of these traditional rowing boats saved him and some colleagues from spending an uncomfortable night on Caher Island near Inishturk in 1910:

> In the evening the wind and sea suddenly got up; our boat after many attempts failed to get alongside the rock at the landing-place; some of the party could not swim, or we might have reached her that way. Rain was clearly not far away, and the wind increased as we wandered around the little island, looking before dark for a sheltered place for sleeping. We found one – already occupied. The sight of an overturned curragh and a face peeping in amazement from underneath it was as welcome as it was unexpected: in a few moments two lobster-men had their boat in the water, and one by one we were safely ferried to our tossing hooker.(WW)

Some curraghs are used on the island today, but now they are powered by outboard engines. Their characteristic raised bows are designed to break through the big Atlantic rollers.

Clare Island

As the powerful ferry pulled away from the pier at Roonagh Quay into the rough Atlantic waters, I could clearly see the great hump-backed outline of

Clare Island across the mouth of Clew Bay. Famous as the ancestral home of the legendary Pirate Queen Grace O'Malley (*Granuaile*), the island's silhouette is dominated by the peak of Knockmore at 462 metres high on the north-western edge and the more gently sloping Knocknaveen in the centre. The western cliffs are among the most dramatic in Europe, and are home to large numbers of nesting seabirds, including a recently established colony of gannets.

I had come here to join a gathering of scientists and naturalists to explore the natural history of the island, to attend the launch of a biography of Praeger by Timothy Collins and to celebrate the Clare Island Survey of 1909–11. The original survey had been organised almost single-handedly by Praeger, who explained his reasons for choosing this location for intensive study:

> The unexpectedly interesting results arising from biological team-work on the island of Lambay near Dublin led to a proposal that a similar survey should be made of one of the western islands, embracing everything connected with its geology, zoology, botany and so on. Clare Island

> was selected, as board and lodging for working parties were possible there, and transport was not too difficult – qualifications which scarcely held for the Great Blasket, a close runner-up. Clare Island had also the advantage of a diversified surface, made up of rocks of various ages, and a hill of 1,520 feet dropping into the Atlantic in a grand cliff covered with alpine plants above and with great bird colonies below.(WW)

Praeger set to, with characteristic energy, organising multiple teams of up to a dozen of the leading experts in their specialist fields in Ireland, Britain and other European countries. They scoured Clare Island, neighbouring islands and the seabed for records of plants and animals and to describe placenames, agriculture, climate, geology and antiquities. In total, a hundred workers took part in the surveys, including many well-known scientists. 'For three consecutive years six or eight parties went out in spring or summer or autumn, and indeed there was no month of the twelve in which one of our collectors might have not been found investigating seaweeds or earthworms or mosses.' His own work in organising the massive undertaking including the publication of the results in three large volumes 'occupied his leisure time for six years'.[45] He found the whole exercise to be 'a full, stimulating and interesting time'. Praeger also found the results rather surprising, 'especially as regards the fauna and flora, in view of the fact that the island and adjoining areas are barren and wind-swept in comparison with lands further east, and much of them covered with unproductive peat-bog':

> Of a total of 5,269 animals observed – ranging from mammals down to microscopic rhizopods – no less than 1,253 species were found to be hitherto unknown in Ireland, of which 343 were unrecorded also from Great Britain, and 109 new to science. Of the 3,219 plants collected, from phanerogams to diatoms, 585 were new to Ireland, 55 new to the British Isles and 11 new to science.(WW)

For its time this was a substantial result, but there were disappointments. Unlike other marine islands throughout the world, no species was found that was endemic to Clare Island, and it was difficult for

Praeger to reach any conclusions about colonisation of the island other than it was an extension of the neighbouring mainland. By this stage, the old pursuits of simple collection of records were beginning to be overtaken by more modern scientific investigations of the ecological processes involved in change in the natural world. Today, the overriding cause of changes is human-induced climate warming and all the attendant impacts on nature.

In retrospect, Praeger's ground-breaking Clare Island Survey is now seen as a thorough baseline measurement of an undeveloped area which has not been immune from the pervasive and long-term effects of habitat loss and land-use change in Ireland and Europe. This was the understanding that led, almost a century later, to the Royal Irish Academy repeating the exercise with the intention of assessing and evaluating change on the island over the intervening years. Over fifteen years, modern teams of specialists used Praeger's baseline to assess the environmental changes that have taken place over the last century on a typical island off the west coast of Ireland. The results have been published by the Academy as a series of thematic books.[46]

This time, the surveys found that some of the

wilder parts of the island appear to be unchanged, but a quarter of the island's wild plants recorded by Praeger have disappeared in the previous century. Some new plants have arrived in the intervening years, but these are mostly garden escapes. Many of the new animals recorded were found among forests of kelp examined using sub-aqua equipment unknown in Praeger's survey, which had to rely on offshore dredging.

The lives of the islanders have also changed radically since then. In Praeger's time, the people were largely self-sufficient, growing and raising most of their own food needs. Today, the farming enterprises, which are dominated by sheep grazing, depend completely on EU support, and many of the young people have left to find better-paid jobs on the mainland. The pre-Famine population of 1,600 was devastated, and the number of residents in 1851 reduced to just 545. By now only around 150 people live here, but almost half of the houses are holiday homes occupied at most for a few months each year. During the long, dark winter there are few lights in the windows.[47]

During my visit I camped on the flat, sandy grass above the beach where Praeger and a group of his eminent colleagues posed for an iconic photograph sitting on a traditional wooden boat in front of Granuaile's castle. Again, Praeger extolled the virtues of the curragh as 'the usual means of communication in the island regions':

> On Clare Island one day, when some heavier boats were anchored close in against the sand in the sheltered bay by the castle, a sudden violent squall came in from the east, raising a sea which threatened to drive them ashore and smash them to pieces. The islanders dashed out from the little harbour in curraghs, and through the breaking waves to the boats, which they towed out to safety. It looked a crazy and dare-devil performance, but to them it was all in the day's work.(WW)

During my three-day stay, I walked to all corners of the island, including the seabird colonies of the northern cliffs and the old abbey in the west with its recently restored wall paintings. The building dates from the 12th century, being rebuilt around 1460, and is unique in Ireland because of the extent of its surviving medieval paintings on walls and ceiling. The abbey offers one of the best opportunities to experience what a medieval painted interior would have looked like. In one of the most stunning clifftop locations in Ireland I passed by the lighthouse, from which I could look out over the rough sea to Achill Island in the north and Clew Bay to the east.

Clew Bay

As I look back into Clew Bay from the summit of Clare Island, the conical peak of Croagh Patrick dominates the scene and below it is laid out a scatter of hundreds of small islands, each one a rounded hill or drumlin formed after the melting of a substantial glacier in the last Ice Age. Some of the islands are joined together by narrow gravel banks that can be topped by the tides. These are similar to the islands and pladdies in Strangford Lough, and both bays hold the best examples of this type of drumlin landscape in Ireland. Lying in the shadow of Croagh Patrick is Bertragh Beach, a long, narrow sand dune spit that links into an island at the northern end. I camped here one night on my way north and walked to the end of the dunes. At one point, the spit is so narrow that it is in danger of being breached by the sea. Yachts entering Westport have to navigate carefully between the islands to avoid running aground on one of these gravel banks.

So far, my sailing experiences around Ireland have all been fairly manageable, but I have the advantages of a depth sounder, good weather forecasting based on satellite imagery, high-precision instruments including a GPS chartplotter as well as the comfort of an ocean-going yacht. In Praeger's time things were largely dependent on the skill and experience of the boat handler. He clearly took risks both on land and sea to reach the remoter areas which held natural history secrets. This passage illustrates that clearly:

> Several islands lie off the South Mayo coast which are well worth a visit by those to whom the heave of the ocean is an exhilaration and not a burden. ... Many a wild crossing I have had among these islands. I remember one from Achill Sound to Clare Island when it was blowing a gale. Our boatman rather begrudgingly agreed to take half his usual complement of passengers but, when he came out from the tail of Achillbeg into the open, he would have turned had it been

> possible. There was nothing for it but to carry on, and away we went under one scrap of sail, over great waves roaring in from the west. The boat rushed down into deep troughs where there was no breath of wind, and only water all round, and up again over high crests ... where the wind half choked us, and we got a momentary glimpse over far-stretching angry seas to distant black foam-rimmed cliffs. Half-way across the waves threatened to comb and one of them spilled a plentiful cascade of white frothy water over us. Down crashed our rag of sail, and we flew down wind under bare poles, while each following sea was a wall of water hovering high over our stern and threatening annihilation.... There was no thought of panic, nor time for it: only a sense of intense exhilaration, coupled with thankfulness that our boatman is experienced and cool-headed.(WW)

I had long held an ambition to cycle along the Great Western Greenway that runs from Westport to Achill Island along the east and north sides of Clew Bay. So I borrowed a bike and set out on this forty-two-kilometre cycling route, one of the first long-distance cycleways in the country, which follows the line of a disused narrow-gauge railway. In the late 19th century Arthur J. Balfour had introduced an Act of Parliament providing State assistance for the construction of light railways to disadvantaged areas in Ireland. The Westport line was extended to Achill Sound in the 1890s, and this became one of the so-called 'Balfour Lines'. Stations were built at Westport, Newport and Mulranny, while the full line to Achill was completed in May 1895. The railway company opened the luxury Mulranny Hotel in 1897, and a combined rail and hotel ticket was available to patrons. The hotel was equipped with every modern convenience of the time, including electric light and, by 1900, hot and cold water baths were also available. The railway became disused in the 1930s with the wide availability of motor cars, and today the embankments and cuttings provide a wonderful, quiet way to see the coastal landscape of this part of Mayo. It was a tent overnight on the Mulranny sand dunes for me instead of a luxury hotel bed.

Achill Island

The Greenway ends at the bridge onto Achill. People from this area have strong links with Scotland, as they frequently went in the past to work on the potato harvest there to make some much-needed cash. At the end of the 19th century it was a remote and poverty-stricken region, with Achill Island being one of the worst-hit by famine and emigration in the 1840s. In the summer of 1894 four Hookers carrying hundreds of islanders set out to sail from Achill to the mainland at Westport, where they were due to meet a steamer bound for Glasgow and the Scottish potato fields. Some people who had walked the forty miles by road were already on board the ship. Many of the young people from the island had never seen a steamer before and, as their vessel, named locally as 'Jack Healy's Hooker', approached the ship, the passengers rushed on deck to one side of the boat to hail their neighbours. Within minutes the hooker heeled over and capsized. Some were thrown into the sea, while others were trapped below deck as the big wooden boat sank to the bottom. In the end, thirty-two people were drowned, twenty-five of them young women and the youngest only twelve years of age. This was the kind of risk that many islanders faced every year, as their connection to the rest of the world depended on taking to the sea.[48]

Praeger wrote that Achill was 'an island only in name for the narrow passage which cuts it off from Curraun is crossed by a substantial bridge. Achill, wind-swept and bare, heavily peat-covered, with great gaunt brown mountains rising here and there, and a wild coast hammered by the Atlantic waves on all sides, but the east, has a strange charm which everyone feels, but none can fully explain.'(WW) My own first visit to Achill Sound was by inflatable boat when surveying the whole west coast for harbour seals. The channel narrows to extensive shallow mudflats that weave in and out of smaller islands, where we found a small group of seals. Our crew landed close to the bridge and took to the nearest pub in Achill Sound for refreshments. Praeger had read about Achill in a book published in 1838 but, he wrote:

> My first glimpse of it at first hand was sixty years later, in 1898, following which, in 1904, I made a careful botanical survey of this highly interesting area. The eagles were gone, save that an occasional itinerant golden eagle was

> reported still; but the flora was unchanged – as it still is – except that an increase in the amount of tillage, due to better drainage, tended to provide a habitat for additional colonists from the east. ... In August the rye stands up white against the bogland tinged purple with heather, with an increasing golden tint as the tufted spikerush assumes its rich autumn colour, among patches of crimson cotton-grass and white mat-grass.(WW)

On the slopes of Slievemore facing south across Keel Bay I visited a deserted village that consists of up to a hundred ruined stone cottages located along a stretch of green road. I spent a while wandering among and inside some of the old houses and through adjacent fields with their lazybed cultivation ridges, imagining what life would have been like for people living here in the early 19th century as they eked out a subsistence lifestyle on the hillside. The whole of Achill was badly hit by the Great Famine of the 1840s and emigration became the only remaining option for many of the surviving villagers.

The old village on Slievemore is the largest and most recently abandoned of several 'booley' settlements on

Achill Island. After the permanent residents left the village it was still used in the summer. The villagers from Pollagh and Dooagh would take their sheep and cattle to graze on the rich mountainside pastures of Slievemore during the summer months. They stayed in the old cottages, returning to their lowland villages for the winter. The practice of booleying, or transhumance, was continued in Achill long after it was abandoned in other parts of Ireland and western Europe. Praeger may well have met some of these booley occupants when he walked all over the hillsides of Achill. Typically, he claimed that, 'the ascent of Slievemore (670 metres) from Dugort is easy, and you obtain a bird's-eye view over the whole island and a vast extent of sea and intricate coast besides'.(WW) He recalled coming down the steep mountainside one stormy day:

> There was half a gale blowing right up the slope from the ocean, and I had to fight my way downhill. I came to one of the little flat platforms that sheep make when they stand for shelter behind a stone. And in the middle of it a fox sat, contemplating the ocean – a fine fellow, with a grand tail and much white about the muzzle. I

approached quietly till I stood only a yard behind him, but the roar of the wind prevented his hearing me, and he kept his head steadily fixed down the slope. I studied him for a long time; then put out my stick and touched him on the back. He was so taken by surprise that he stared at me for quite two seconds before making off with remarkable speed.(PS)

I went for a swim at Keem Bay at the very western tip of Achill. This is a cliff-bound cove with a beautiful sandy beach at the head. Throughout the early 20th century this was the location for the capture and killing of basking sharks, the largest fish in the Atlantic. Kenneth McNally, by profession a photographer with Ulster Television, filmed the fishery and wrote about it in his book *The Sunfish Hunt*.[49] A shortage of fuel after World War II led to an increasing market for shark oil for use in certain industrial products. The slow-moving sharks swam into the bay to feed on dense swarms of plankton near the sea surface. Here they became entangled in nets set by the islanders, who then launched their lightweight curraghs and killed the struggling fish, stabbing them with scythe blades attached to long poles. Over the thirty-year period up to the 1970s more than 12,000 basking sharks were landed on Achill – an average of at least 400 fish per year. These harmless sharks were caught at a number of regular haunts along the west coast in the preceding century and large quantities of the valuable oil were exported to England. Not surprisingly, catches declined markedly towards the end of this period and, with the availability of alternative mineral oils, the market for shark oil disappeared, allowing the few remaining animals to survive.

Today, the sharks are back, their population slowly recovering from this classic example of overfishing. Jane Clarke has written a moving poem about this experience called 'At Purteen Harbour':

At Purteen Harbour

basking sharks, docile as seal pups,
harpooned and netted from currachs,

were towed one by one to the fishery
at the slipway. Fathers and sons

sliced off dorsal fins and hacked
through blubber to reach oil-filled livers.

Sweating in burn-house heat,
they shovelled bleeding flesh

into the rendering machine.
They couldn't wash the smell

from their skin, not if they swam
to Inis Gealbhan at the end of every shift.

Year by year the catch grew less,
then disappeared.

But late last April, old men cheered
from the headland, and said,

It was as if they'd been forgiven—
a school of twelve cruised into Keem Bay,

moon tails swishing, fins proud
as yawl sails above the waves.[50]

Elsewhere I have watched a basking shark surface beside my boat, lazily cruising through the rich waters. They are truly impressive fish, completely docile and harmless, unlike some of their relatives.

Bills' Rocks

Although I have never managed to visit them, the Bills' Rocks, some ten kilometres south-west of Achill, must be among the most remote parts of the Irish coastline. These are three rugged rocks rising almost vertically from the Atlantic. The Bills' Rocks were called after the Danish Captain Mathias De Bille. In 1781, the Royal Danish Navy frigate *Bornholm* left Copenhagen bound for the Danish West Indies. The following January, as the ship reached the Irish coast, a hurricane drove it south towards Clew Bay. The foremast and bowsprit were lost but the captain managed to steer the ship into Melcombe Bay near Newport to avoid the vessel being smashed against the Bills' Rocks.

Praeger clearly did visit these isolated rocks, measuring just over one hectare in total area, probably as part of the Clare Island Survey in 1909–11. He wrote:

> I remember a lovely belated spring display of sea-pink and scurvy-grass on the Bills, those lonely stacks away to the south-west of Achill. It is not easy to land here, for on the calmest day the almost imperceptible heave of the Atlantic

> changes to foaming breakers against the rocks. One chooses a vertical rock-face (with a sufficiency of foot-holds and hand-grips), for there the wave goes straight up and down, and there is no danger of a capsize. Then one notes a niche for one's foot, and steps ashore on the top of a swell. The return is mostly more difficult; sometimes the driest way is to throw one's clothes into the swaying boat and swim for it. Swimming, indeed, is the primitive and proper way of approaching an island.(BS)

The Bills' Rocks have held a fascination for ornithologists for over a century, but there have been few attempts to survey the birds here. David Cabot has made five visits over a period of fifty years, with the most recent in 2018, when he was the first person known to camp here overnight. Just four seabird species were definitely recorded breeding, and there was a tantalising suggestion that a few Leach's petrels, one of the rarest birds known to breed in Ireland, were nesting here.[51] They are extremely difficult to observe, as they nest deep in underground burrows. Five years earlier, Eoin McGreal had confirmed their presence by playing the bird's call on his mobile phone outside a nesting burrow and eliciting a response from an incubating bird.[52]

Blacksod Bay and Broadhaven

While surveying seals on the west coast I was on a large inflatable boat as we passed through the town of Belmullet via a narrow canal linking Blacksod Bay to Broadhaven. In 1715 Sir Arthur Shaen began building the town at Belmullet on a wet and marshy area that linked the Mullet Peninsula to the mainland of Mayo. To drain this area and to form a passageway for small boats between the two nearby bays, Shaen had a canal excavated and a sluice was erected at the Blacksod Bay side to allow boat traffic to and from the Mullet Peninsula to pass along the shore. Development of the town proved to be a slow process, and the canal fell into disrepair by the mid-1700s. A century later work began to restore the canal, but it was not completed until 1851 due to the Great Famine, which had a particularly devastating effect on the Erris region. A report in that year stated that the canal was then being used extensively and also mentioned the use of a swivel bridge. This has been replaced by a modern road bridge, while the original footbridge was demolished in 1989 and replaced by a cable-stay bridge named, appropriately enough, 'Reconnections'.

The shallow, sandy waters of Blacksod Bay provide ideal seabed habitats for the specialised communities of maërl (algae that create a hard calcareous structure) and seagrass meadows similar to those found in some other west coast bays. On the west side of Blacksod Bay there are also significant submarine reefs constructed by marine worms known as Serpulids. These habitats are so rare in Europe that the whole bay has been designated as a Special Area of Conservation. There are currently problems here with unsustainable fisheries. Reports by the Marine Institute and the NPWS conclude that scallop dredging is incompatible with maintenance of maërl and seagrass communities, and may significantly impact reef fauna, including Serpulid and Laminaria in Clew Bay and Blacksod Bay. Impacts of scallop dredging in sedimentary habitats may be significant in Clew Bay, Broadhaven Bay and Blacksod Bay given the spatial extent of the fishery and the protracted fishing season. The seasonal oyster fishery will add cumulatively to this effect. The marine worm-dominated community complex in Blacksod

Bay, which had previously (2008) been shown to be comprised of large aggregations of biogenic reef formed by *Serpula*, now consists of broken tubes. Very few living aggregations are still present, and the total habitat area of this marine community type has been negatively impacted. The cause of this impact is physical damage due to benthic dredging.[53]

The Irish Wildlife Trust pointed out that the NPWS report also found maërl habitat to be damaged, even though the Marine Institute recommended this should be avoided by the bottom fishery. The protected sedimentary habitats in Blacksod Bay are frequently dredged, and this is adversely altering these protected ecosystems despite legal protection.[54] The marine biologist Bob Brown says that it even destroys the hydroids upon which scallops depend for larval settlement.

Passing through Belmullet, the wide shallow bay of Broadhaven stretches away to the north. Praeger remembered a trip here, probably during the 1914–18 period.

> The first time I saw Broad Haven was before sunrise on a lovely morning in May many years ago, when as a war correspondent, I joined a

> landing party of a mate and two seamen from the Congested Districts Board steamer Granuaile. Their fell purpose was to capture and forcibly abduct a Spanish jackass which was roaming the hills 'somewhere in Mayo'. We landed on a stony shore where a couple of cottages gleamed white against miles of bog, and for a while I was left alone in the company of some ducks and three contemplative goats, who watched with profound attention the first sunlight gilding the cliffs of Erris Head out to the west. ... I think the Congested Districts Board had a number of these tall Spanish beasts on the west coast for a time, in an endeavour to improve the stamina of the local donkey population.(WW)

The Congested Districts Board for Ireland was established by the British government in 1891 to alleviate poverty and cramped living conditions in the west and north-west of Ireland. Devastated by the Great Famine and the mass emigration that followed, the remaining people in these areas were still extremely poor, and there was much political pressure for change. The government response was based on a Conservative policy of 'Constructive Unionism' or 'killing Home Rule with kindness'. The Board was set up to alleviate poverty by paying for public works such as building houses, constructing new piers, assisting the development of fishing, modernising farming methods and sponsoring local industries, such as the knitwear business on the Aran Islands, to give employment and stem the continuing emigration from Ireland. Many of its constructions, such as the hundreds of small stone piers on the west coast, remain to be seen today.

Apart from its physical works the Board had a fundamental impact on land ownership, especially on some of the offshore islands. For example, extreme poverty on Clare Island forced many evictions, so the Board purchased the entire island from the landlords in 1895 and resold it to the tenants five years later. The Board was wound up in 1923 by the new government of the Irish Free State and its functions were absorbed into the new Department of Agriculture. However, the historian Diarmaid Ferriter believes 'the bald reality and deep irony was that more was done for the islands under British rule than was done in the early decades of native rule'.[55]

The Mullet

When I first started to work for the Irish Wildbird Conservancy (now BirdWatch Ireland) I set out to visit all of the sites where the organisation had undertaken conservation work. On a fine summer's evening in the early 1980s I arrived on the Mullet, about as far west as it is possible to go on the wild Atlantic landscape of Mayo. Praeger knew the area too, and he gave a colourful description:

> Belmullet has as a stopping-place the advantage of allowing easy access to The Mullet, that remote peninsula, half bog, half sand, full of bays and queer lakes and outlying islets; treeless, sodden, storm-swept and everywhere pounded by the besieging sea; beautiful on a fine day, with wondrous colour over land and ocean, desolate beyond words when the Atlantic rain drives across the shelterless surface.(WW)

I camped in a field near the coast, west of Belmullet, and close to a small wetland at Annagh Marsh. The grassland around me was rich with wildflowers, including sheets of yellow bird's-foot trefoil and purple seaside pansy. As I cooked up a meal in the setting sun, I could hear corncrakes belting out their electric calls into the warm air and a snipe drumming overhead. Even a century later this was still clearly a magic place. When I wandered around the marsh early next morning, I spotted an unusual-looking wader which, unlike most related species, was actually swimming in circles on the shallow pools. Praeger wrote:

> On the edge of the Atlantic, in a place that shall be nameless, my wife and I had the most interesting ornithological adventure that has befallen us. We found ourselves suddenly among fairy-like birds, quite unknown to us, but evidently belonging to the Plover family. The most extraordinary thing was that they displayed absolutely no fear of us, darting around our feet, running over the slender water-plants which filled the pools and which, one might have thought, would scarcely have supported even a dragonfly, uttering even a small sharp cry. It was only when we got back to Dublin that we found that we had stumbled on the sole Irish breeding-haunt of the Red-necked Phalarope, which had been discovered only the previous year.(WW)

For most of the 20th century, Annagh Marsh was the only regular breeding site in Ireland (and the most southerly breeding site in the world) for red-necked phalarope. These small arctic waders are distinguished by the bright red neck colour of the females. The males are more camouflaged, as they take on the role of 'mother' by incubating the eggs and rearing the chicks. First discovered here in June 1900, there were various estimates of up to fifty pairs breeding until 1929. Thereafter they started to decline, despite the fact that the location of this breeding place remained confidential. This was thought to be because of pressures from egg collectors, along with the trampling of nests by cattle. In 1937 the Irish Society for the Protection of Birds acquired the ground and fenced the area to restrict access and cattle grazing. By 1968 this site held between three and five pairs, along with twenty non-breeding birds. While the sanctuary at Annagh had been protected for many decades, the decline in numbers of the birds continued, and it was recognised that the pools were becoming overgrown with vegetation. Local volunteers organised several work parties to create more open water, which the birds need for foraging. Unfortunately, the historical decline in the breeding population continued, and the last documented breeding was in 1980, with only occasional birds being seen over the next ten or so years.

More recent management of the area for other important breeding wader populations has resulted in the return of red-necked phalaropes to breed each year since 2015, as well as an increase in the numbers of breeding lapwings. Between 2002 and 2005, funding support from an EU LIFE-Nature programme enabled BirdWatch Ireland to restart active management and to graze up to 156 hectares surrounding the nearby Termoncarragh Lake through management agreements with farmers and direct land acquisition.

In addition to Annagh Marsh, the reserve at Termoncarragh is an important area for grassland biodiversity. The area has generally been enhanced and managed as species-rich, semi-improved grasslands, while the coastal heath and grassland have been maintained as such. The management of the meadows is primarily focused on breeding corncrakes, with up to two calling males holding territories in most years, while the coastal grasslands are managed for wintering geese. This area has attracted, on occasions, up to a thousand wintering barnacle geese. These are part of the bigger flock that flies daily from the offshore islands to feed on the grasslands adjacent to Termoncarragh Lake. However, other species of importance have benefitted too, including breeding skylark and meadow pipit. The meadows have also become important for rare invertebrates such as the great yellow bumblebee.

These wet grasslands are nationally important for the suite of species mentioned, and there are multiple threats to them. Dave Suddaby, BirdWatch Ireland's Reserves Manager says: 'Our knowledge of the impacts is ever-increasing but we still need to have a better understanding of these in order to manage the site. The whole ecosystem, including the birds, insects, vegetation and the farmers associated with the site, need to be taken into account to enable them to flourish into the future.'

Inishkea Islands

The Mullet and nearby sites on the Mayo coast were among the areas of sand dune machair that held the largest numbers of breeding waders when I undertook a full survey of these birds from Donegal to Galway in the mid-1980s.[56] Noisy pairs of lapwings circled overhead, snipe chipped from the wetter areas and redshanks perched on fence posts between the grazing cattle. I remember finding a clutch of eggs laid by a dunlin in the hoof print of one of these beasts. But all this was to change in following years. Dave Suddaby has led the most recent breeding wader surveys on these western coasts and he has recorded dramatic declines of seventy to eighty per cent in most of the species. Among the last coastal areas to still hold significant numbers of waders in summer are the remote Inishkea Islands about five kilometres west of the Mullet.

These islands are almost frozen in time. In the 19th century, each of the two main islands had a busy village. I have been here many times over the last few decades, usually to help out with bird research projects led by David Cabot, who has been visiting the islands since 1961. Walking down one of the old grassy village 'streets' between rows of ruined cottages I could imagine the children, a century ago, playing among the wildflowers, women working in the potato fields and men hauling their curraghs up the beach below. Surprisingly, the population of the two islands grew over the period of the Great Famine in the 1840s, reaching a peak of 319 people in 1861 at a time when there was a huge decline on the mainland due to disease, death and emigration. While numbers living on the South Island started to decline in the early 20th century, the North Island village was growing, and had its own school, a police barracks and even a pub.[57]

To reach the Inishkeas I drove to the very southern tip of the Mullet at Blacksod, where I met the rest of the team and we boarded a small fishing boat to take us out across the sea to the substantial pier on the South Island. Sometimes, the alternative was to launch an inflatable boat and risk the big Atlantic swell and currents around the headlands. Just off the island pier is the small islet of Rusheen, which is linked to the South Island at low tide. Walking around the rocky shore I noticed large pieces of rusting ironwork, the remains of heavy machinery and fallen brick walls lying about the island. This was the site of a whaling station in the early years of the 20th century. A Norwegian company had bought the site on the islands from the Congested Districts Board, whose chairman, W.S. Green, had championed the whaling industry as a source of much-needed employment on the west coast. In 1908 the station was built from a shipload of imported timber, a slipway and flensing stage (for cutting up the whale meat) were built and three large winches were installed to haul the whale carcasses from the sea. Over 200 fin whales were caught and processed here in the next five years, as well as smaller numbers of blue, sei, sperm, right and humpback whales. However, declining catches and the onset of the First World War caused the closure of the whaling station, and a similar one on the mainland at Blacksod, after a few years.[58]

After this extraordinary episode in their history, life for the islanders returned much to normal. Fishing was the main source of income, and the total number of islanders remained around 300 until the 1920s, when disaster struck. Without warning, a huge storm hit the west coast in October 1927, and ten of the young men from the islands were caught out in their small fishing boats and drowned. The same sudden storm on the west coast, known as 'the Cleggan disaster', killed thirty-seven more fishermen from different harbours on the west coast. The remaining population of the Inishkeas was subsequently relocated, at their own request, to houses on the mainland, from where they could see their former homes.

This very isolation and lack of disturbance during the colder months is the principal reason why the

Inishkeas host a large number of wintering birds, notably flocks of up to 5,000 barnacle geese that fly in during November each year from their summer breeding grounds in East Greenland. With David Cabot, Michael Viney and Steve Newton, I was fortunate to visit the arctic end of this long migration route in the 1980s in support of David's long-term studies of the geese. The Inishkea Islands are like an international airport hub for geese, as about half move quickly on to other coastal areas in Ireland, while the remainder spend the winter here. He has renovated one of the small stone cottages on the island, from where he watches the flocks through a powerful telescope. A sample of the birds are caught and marked with numbered leg rings, each of which has a unique identifying code, similar to that on a vehicle registration plate. From this information David has built up a detailed picture of these geese as they stay together in families throughout the winter. He notes all this information in a set of unique journals, which detail a lifetime's knowledge of the islands and their natural history.

The reduction in hunting pressure from the 1970s onwards led to a recovery in the population of grey seals on these uninhabited islands. The animals assemble in the autumn and the young are mainly born in the months of September and October. The white-coated pups are unable to swim for the first few weeks of life, and lie around among the boulders moulting and waiting for occasional feeds when their mothers come ashore to suckle them. I was able to scramble close enough to one pup to stare into its large watery eyes while it watched me carefully. Its mother was not far away, waiting in the surf for me to retreat. Up to a thousand of these seals gather on the islands in winter, when the local breeders are supplemented by incoming animals from Scottish breeding colonies.

North Mayo

Leaving the long inlet of Broadhaven and rounding Benwee Head, the coast turns directly east, and from here I descended to the isolated settlement at Portacloy, where curraghs were still stacked upside down just above the beach. Caesar Otway, in his *Sketches in Erris and Tirawley*, gives an account of exploring the north Mayo coast around Portacloy in what was presumably a large curragh. Otway was a man of great energies and curiosity, though he was not shy of saying what he thought of the poor natives in the 18th century:

> I consider the coast-guard establishment one of the greatest blessings ever conferred on Ireland—a positive blessing in not only putting an effectual stop to smuggling, the nurse of profligacy and crime; to wrecking, the stimulus to dishonesty and cruelty, all-around our shores; but also, in locating prudent, honest, humanized, and often religious men, with their nice wives and children, and all their clean and decent habits amongst a dirty, ignorant, and careless people.[59]

Otway was completely overawed by the cliff scenery of north Mayo. The naturalist Richard Ussher was equally impressed by these cliffs many years later, when he was searching for sea eagles as he prepared his landmark book on *The Birds of Ireland* with Robert Warren.[60]

Just two kilometres off this coast is a small group of five cliff-bound rocky islets, with the highest rising to almost 100 metres above sea level. The sea has cut a tunnel entirely through one of the rocks, *An Teach Beag*. The Stags of Broadhaven are a popular site for visiting divers, sub-aqua teams and kayakers. The historian David Gange landed here during his epic sea kayaking voyage around the Atlantic fringe of Britain and Ireland. 'We reached the Stags with tides running south and swell driving north. This stirred a vortex of sea like a mile-wide whirlpool round the rocks. I ploughed into the rocks, seeking respite, and scrambled up the cliff .... I remember breathing far more heavily than the level of exercise merited, my chest tight with nervous energy.'[61] These rocks were one of the few places in Ireland thought to hold a breeding colony of the rare Leach's petrel. Several of these elusive seabirds were seen and heard by ornithologists who managed to land on the rocks in 1946 and 1947. They were not surveyed again until 1982, when the petrels were found in burrows, a number were trapped and an estimate of at least 200 breeding pairs was made.[62]

To the east near Belderrig, the extensive blanket bog runs almost to the top of the cliffs, and here I called into the modern visitor centre at the Ceide Fields. In the Neolithic (New Stone Age) period this landscape would have looked entirely different, as it supported a thriving farming community. The secrets beneath the bog have been uncovered over the last few decades by archaeologists who carried out a painstaking process of probing the soft peat with long metal rods. These slip easily through the surface layers of the bog, but when they hit a line of stones beneath the peat, a map of the subsurface landscape can be produced. These stones were cleared from fields by Neolithic farmers and built into walls similar to those in the farms throughout western Ireland today. During my visit I followed the archaeologist Seamus Caulfield around the path as he explained how he and his colleagues found this ancient farming landscape of field walls stretching over at least 1,000 hectares. Pollen remains from the bog and the dating of ancient pine stumps at the base of the peat showed that the building of the walls coincided with a period of farming around 4,500 years before the present.

Sligo

Stopping briefly at several more spectacular cliff sites, including Illaunmaister and Downpatrick Head, I reached the wide inlet of Killala Bay, where I crossed the River Moy at Ballina.

Just above the rocky shoreline at Enniscrone on the east side of Killala Bay I found the traditional seaweed baths that have been run by generations of the Kilcullen family since 1912. Praeger would have passed by this building in his ramblings, as bathhouses came to Ireland during the Edwardian era, with about 300 facilities scattered around the country. Nine of those were in County Sligo. Edward and Christine Kilcullen are the fifth generation to own this bathhouse, which replaced a much older establishment from the middle of the 19th century. Inside, original fittings such as the enormous, glazed porcelain baths, solid brass taps and panelled wooden shower cisterns, give the place its unique historical atmosphere. The practice of immersing the body in steaming hot water filled to the brim with fresh seaweed from the local shores is said to provide relief from painful symptoms of rheumatism and arthritis. These therapeutic properties are attributed to the high concentration of iodine that occurs naturally in seawater, and the brown seaweed concentrates it in its fronds. As I sank into the silky, aromatic water I became totally relaxed and refreshed for another day's adventure.

I spent a year studying the water birds of Sligo Bay, during which time I gained a detailed knowledge of all the laneways that lead down to the shore. Walking the sandy beaches and rocky points gave me a wonderful opportunity to explore this complex stretch of coastline. Praeger also visited this coast:

> Sligo Bay, which enters the land between Roskeeragh Point and Aughris Head, breaks at once into three long fingers of complicated outline, with points and islands, making an extensive shallow area of calm water invading the limestone country, with miles of bare sand at low tide. The southern finger runs up to Ballysadare. (WW)

At low tide, Ballysadare Bay is a maze of shallow channels and sand banks. One of these is marked on Ordnance Survey maps as 'The Great Seal Bank', and I imagine that it was once a place where local people hunted the seals, hauling their quarry back to the shore in small boats. I was here in the 1970s as part of a team of biologists undertaking the first complete census of harbour seals in the island of Ireland.[63] Travelling in two fast inflatable boats, we surprised hundreds of the animals hauled out on the sand banks. They shuffled down to the water and, with typical curiosity, swam around our boats at a safe distance, puzzling at the strange human beings. The main seal herd sometimes left behind some young pups, as they were not able to move so quickly into the water. Occasionally, all the seals had already entered the water, having heard our engines before we saw them, and we were limited to counting their distinctive tracks on the sand.

The mouth of the central bay is largely filled by the flat Coney Island. When the tide was low I walked out across the sand towards the island following a line of marker posts. Several vehicles passed me as I walked, rushing to beat the rising tide. It is a long walk across a kilometre of hard sand, and I was anxious that I would not be stuck on the island for a full cycle of the tide. At the northern edge of the bay a channel has been dredged to allow small cargo vessels to reach the quay walls in Sligo Town. It is also used by leisure boats that tie up in the mouth of the Garavogue River.

North of Rosses Point lies the third inlet of Drumcliff Bay, best known for the historic house of Lissadell, which was a favourite holiday destination of the poet W.B. Yeats and birthplace of the revolutionary Countess Markievicz, born Constance Georgine Gore-Booth. Her leadership role in the Easter Rebellion of 1916 and the subsequent struggle for freedom in Ireland led to her arrest but, because she was a woman, she narrowly avoided execution. I joined a guided tour of the house and was amazed to see that it remained much as it would have been at the time of the Gore-Booth family. Sir Henry Gore-Booth, father of Constance, was an adventurer who sailed his own yacht *Kara* to the Arctic where, in 1882, he was involved in the rescue of a fellow Arctic explorer called Benjamin Leigh Smith. He made several voyages sailing to Norway for salmon fishing, to Spitzbergen Island, to Nova Zembla in arctic Canada and to northern Greenland over a period of twenty years. He was keen to see a whale harpooned and asked that he be called as soon as one was sighted. The entry for his log on 27 May 1884 records 'the report of the gun awoke me. The noise of the line running out fetched me out of my berth, and the watch yelling out "A fall! A fall!" caught me with one leg only in my trousers'. Sir Henry's library of books from the *Kara* is on display at Lissadell, together with a stuffed polar bear he brought home from the Arctic and a 19th-century model of a rigged yacht. Sir Henry was also a prolific writer on a variety of topics, including Arctic exploration, yachting, whaling, polar bear hunting and shark fishing.

At home, his lordship must have been reminded of the arctic each winter where, on a large field right next to the wooded boundary of Lissadell, I watched a huge flock of barnacle geese, peacefully grazing on the grass that is kept short to attract them to this sanctuary. These handsome black-and-white birds breed in the frozen wilderness of east Greenland in summer, arriving here in October after a long migration via Iceland. As the flocks moved across the field, I could hear a constant low murmur as the geese carried on a conversation with each other. The goslings of the previous summer stay in close family groups with their parents through the winter, including the return trip to Greenland the following spring. This flock that moves around Sligo is the largest concentration of these geese in Ireland, with over 4,400 recorded in the latest census in 2018.[64]

North of Lissadell, the coast straightens out and breaks into a series of beaches and headlands. The shallow sandy shoreline here proved to be a disaster for no less than three of the tall ships that were part of the Spanish Armada. Hundreds of soldiers and seamen were drowned on the beach at Streedagh in 1588 when their ships were driven ashore and battered to pieces in a gale. Most of the survivors who reached the shore were beaten and robbed by the local population, with any remaining Spaniards being brutally murdered when the English garrison arrived from Sligo. One of the officers later reported, 'I rode along that strand .... there lay a great store of timber of the wrecked ships .... in one strand lay eleven hundred corpses'.[65] One of the few to escape was Captain Francisco de Cuellar, who recorded that wolves were seen on the shore eating some of the bodies. He walked inland as far as Lough Melvin, where he and a few companions were hidden by the Clancy family on a lake island at Rosclougher. He eventually made his way back to Madrid, where his memoirs were found much later and form the main source of information on the disaster at Streedagh.

Today the long sandy beach at Streedagh is one of a number of popular surfing waters in the north-west coast of Ireland. At the back of the sand dunes were parked lines of distinctive camper vans with surfboards strapped to the roofs. As I walked along the beach, lines of rollers were driven ashore in a strong westerly wind. I reached the stepped cliffs at the end of the beach, where I found an abundance of fossils preserved in the limestone. Large single corals the size of a wine bottle and massive colonies of tiny coral tubes appeared white against the grey stone. Many of these can be found in the large boulders that make up the huge storm beach at the top of the sand.

Inishmurray

Looking offshore from the beach at Streedagh I could see the flat island of Inishmurray, marked by a line of breakers along its shoreline. The island is best known for its remarkably well-preserved monastic remains, which date from the early medieval period. Founded here in the 6th century, the monastery was raided by Vikings in 795AD, suggesting that it had valuables worth stealing. The main feature today is a massive circular stone enclosure protecting a number of churches and beehive cells. It is now thought that these were an important destination for early pilgrims coming to worship at the tomb of Saint Molaise.[66]

Praeger, who first visited the island in 1896, wrote:

> Right opposite to Mullaghmore, four miles out lies Inishmurray. It is a low islet a couple of hundred acres in area, much exposed to Atlantic storms; yet the place is full of early Christian antiquities, pointing to considerable importance in old days; and it still has a population numerous in relation to its size and agricultural possibilities, largely dependent on fishing for its livelihood.(WW)

The people of Inishmurray also had another trade that kept them in business. The distilling of illicit whiskey, known as *poitín*, was carried on here, out of reach of the strong arm of the law. Barley was grown on the island as one of the main ingredients. The distilled alcohol was stored in wooden kegs and hidden on the island until it could be transported under cover of darkness for sale on the mainland. Occasional raids by the authorities achieved little, as the approaching boats could be clearly seen and the islanders quickly hid the delicate equipment and kegs of whiskey among the extensive boulders on the shoreline. According to Dr Patrick Heraughty, an islander who left to study medicine and later wrote an account of the island, 'there were 102 people living in fifteen houses'. World War II brought serious difficulties for the island people. Wartime fuel shortages led to the end of a motorboat service bringing supplies to the island. Sugar, one of the main ingredients in the manufacture of *poitín*, was rationed, and it became impossible to obtain enough to continue making the valuable product. Emigration increased steeply as islanders left to find a new life in Scotland or America. As island life became more difficult, evacuation of the last six families, with a total of forty-six people, became the only option left in 1948.[67]

As so often happened on the offshore islands, the departure of the human population led to an increase in wildlife. In July 1962, David Cabot spent five days on the island observing birds, and he found a healthy population of eider ducks and gulls.[68] Since then the eider numbers have reached about 100 breeding pairs, making this one of the most important concentrations of the species in the island of Ireland. These ducks were first recorded breeding in Ireland in 1912, so they were probably present when the last of the islanders lived here. In Iceland, where they are common, the people collected the soft feathers lining the nests when breeding was finished and these were sold as eiderdown, best known for its use in pillows and duvets. Perhaps more important is the colony of storm petrels found to be nesting here in crevices in the many stone walls. These tiny seabirds are active mainly at night to avoid predation by large gulls, so counting them is extremely difficult. Nevertheless, in 1979 it was estimated that there were 500 pairs. A ringing study of the birds has shown that there are many interchanges with other breeding colonies on the west coast, as well as with those in the Irish Sea.[69]

Donegal Bay

Donegal Bay is an enormous inlet that stretches from north Mayo to the south-western extremity of County Donegal, dominated by the spectacular cliffs of Slieve League. I called into the Smuggler's Creek Inn, a hostelry founded in 1907 above the cliffs at Rossnowlagh. Here I enjoyed a wonderful seafood meal while gazing out the front windows at dozens of surfers on the wide beach below. The inn gets its name because pirates and smugglers used the creek below the inn to land their vessels and haul their contraband up an old hidden path. To the north of Rosnowlagh I explored the shores of Inner Donegal Bay, the almost land-locked estuary of the River Eske, which lies behind a set of sand dunes at Murvagh. This sheltered sandy inlet is used by a group of harbour seals, which can be seen easily from various vantage points on the top of low drumlin hills.

Sticking out in the centre of Donegal Bay is the long, narrow arm of St John's Point, the end of which is marked by a lighthouse. This point is permanently fixed in my memory, for the only time I rounded it by sea was during a survey of the seals, when our powerful inflatable boat met enormous waves here. Amid near gale force winds we rose up to the crests and slapped down the other side with huge Atlantic breakers rolling in behind us from the west. Marine biologist Bob Brown described the waters at the point as 'one of the most beautiful and striking dives I ever had, outside of the Caribbean'. His description is stunning. 'A dive here reveals a spectacular garden of marine life, thriving in the productive and mobile waters. Most dramatic are colonies of jewel anemones, brilliant pinks, purples, greens and yellows, in patches or clones generated by prolific asexual budding. Because the water here is very clear with good light penetration, the scene is breathtaking with outcrops bursting with psychedelic colours.'[70] I visited St John's Point myself in beautiful summer weather and walked among wildflowers with wonderful unbroken views on all sides from the north Mayo cliffs to Ben Bulben in Sligo and Slieve League in west Donegal. Praeger wrote of this place that,

> From Killybegs also you can visit St. John's Point, which forms the extremity of a narrow five-mile promontory jutting far into Donegal Bay, and differing from all the land to the westward in that it is made of limestone. It is indeed the last outpost of the Carboniferous limestone which

> extends southward across Sligo and the whole Central Plain of Ireland: and it is interesting to note that a few of the characteristic plants of the limestone, such as the Bloody Crane's-bill, the Northern Bedstraw and the Blue Moor-grass, have followed their favourite rock to this remote spot.(WW)

In one respect, Praeger made a rare mistake in his observations here. I know this because I spent a summer on this coast as a university student mapping the limestone rocks, which actually stretch well beyond Killybegs and westward along the coastline to Muckross Head near Kilcar. But the wildflowers near the lighthouse at St John's Point are still just as Praeger found them. The limestone in places forms classic bare pavements broken into slabs just like that of the Burren in County Clare. Here, among the clints and grikes, I found such beautiful plants as hart's-tongue fern, devil's-bit scabious, eyebright and harebells. Wild mountain hares scattered across the unfenced grassland as I strolled along, and the few bushes of hawthorn and willow were bent almost horizontal by constant strong winds from the ocean.

The western extremity of the south Donegal coast is marked by a line of high cliffs that Praeger also knew well:

> The glory of this promontory is Slieve League on the southern shore. A tall mountain of nearly 2000 feet, precipitous on its northern side, has been devoured by the sea till the southern face forms a precipous likewise, descending in this side right into the Atlantic from the long knife-edge which forms the summit. The traverse of this ridge, the 'One Man's Path' is one of the most remarkable walks to be found in Ireland – not actually dangerous, but needing a good head and careful progress on a stormy day.(WW)

It was a beautiful calm day in summer when I took to the sea in a small boat from the pier at Teelin at the eastern end of the cliffs. As the boat rounded the point at Carrigan Head a stunning view of the cliffs opened up above me, echoing Praeger's words. 'The quartzite and gneiss of which it is composed is singularly variegated, and on a bright day the kaleidoscope of colour – yellow, red, white, gold – which it displays is wonderful against the blue of sea and sky.' He was fascinated by the collection of alpine plants which he found on steep slopes of rock here, including alpine meadow-rue, mountain avens and purple saxifrage.

Glencolmcille

Travelling west from Slieve League across wide-open spaces of bog and mountain grassland, I reached the small village of Glencolmcille, set in its own valley facing the Atlantic. Many people, both Irish and international, come here to learn or improve their Irish conversation at classes run by *Oideas Gael.* When I joined a group of geographers and archaeologists in the village, we climbed down into a stone-lined souterrain in the grounds of a church. These underground passages were built in the Early Christian period in Ireland as hiding places from invaders, and may also have been used as safe places to store religious valuables. Praeger advised:

> If you want to visit one of the finest stretches of coast in Ireland, stay at Glencolmcille, and explore not only the Slieve League side, but also the northern shore towards and past Slieveatooey.

> Glencolmcille itself is a lonely and lovely spot, with interesting associations, for hither, in the sixth century, St. Columba journeyed over miles of moorland with a band of disciples, to find in this sequestered valley, shut in between the mountains and the ocean, a spot meet for meditation and prayer. The place is still full of relics of subsequent religious occupation.(WW)

About fifteen kilometres north of Glencolmcille, at Port or *An Port*, there are the remains of a 'ghost village' (also called The Deserted Village), where a number of houses were abandoned during the famines of 1845–52. Of course, millions of people died or emigrated from Ireland during this period, but the houses at Port were never reoccupied, probably because of the remoteness of the place. The houses were all built of local rock, and I noticed the distinctive long stones protruding from under the eaves which were used to tie on the thatch and prevent it lifting in westerly gales. Standing here among the ruins in this remote spot, it was hard to believe that a community of people lived and farmed here some two centuries ago, where huge rocks and boulders seem to be more common than soil.

In 1576 the daughter of Tarlach Neill, head of the O'Boyle Clan, drowned here. In the *Annals of the Four Masters* it is recorded that she had drowned while learning to swim in the river that runs into the sea at Port. However, today this is merely a stream, and an alternative explanation seems more likely. It is suggested that, as she was being forced into an arranged marriage, she ran away to Port followed by the man she was to marry, and here he drowned her. Today there is a slipway beside the rocky beach and a little bridge over the river. It is very peaceful and a place to reflect on the sad events that took place there over the centuries.

Sheskinmore

North of the high Glencolmcille Peninsula the coastline drops dramatically to sandy beaches at Loughros More Bay, where saltmarsh and sand dunes are the dominant features. The beaches and dunes here seem to change almost beyond recognition from one decade to the next as sand moves around, eroding the dunes on one side of the bay and building new ones on the other. Perhaps due to its remoteness and reputation for wind and rain, this is always a quiet place to visit. Even at the height of summer it is possible to be alone on some of the best beaches in the country. Since the 1980s I have spent numerous family holidays in this area and explored its many hidden wildlife jewels.[71] In the middle of the dunes, a shallow lake near Ardara was the breeding place of waders such as lapwing and dunlin nesting in undisturbed marshy vegetation between the sand dunes and the reeds. In winter, Greenland white-fronted geese grazed on the lakeside pastures. The name of this area, *An Seascann Mór*, means literally 'the big marsh' in Irish. In the early 1980s news emerged that the lake was being drained by a local farmer who wanted to improve his lands upstream. Apart from the environmental damage, the neighbouring landowners were not happy, as they found that their cattle could walk around the ends of long fences and into other properties.

The Donegal naturalist, Ralph Sheppard, who knew the site well, alerted the Irish Wildbird Conservancy (now BirdWatch Ireland). Emergency action was called for if the site was to be saved. With the agreement of neighbouring landowners, the IWC installed a simple sluice on the outlet stream, thus controlling the levels of the lake. In the following years

Ralph and I undertook breeding wader surveys here and found a diversity unlike other west coast areas.[72] By the end of the 1980s, the National Parks and Wildlife Service had purchased a sizeable area of the surrounding land, thus securing its future as a natural area, and John Hennigan, the local Wildlife Ranger at that time, became its guardian. The original sluice was later replaced by the NPWS with a more substantial and durable structure, allowing the water levels to be controlled for roosting geese and protected plants.

Today the marshy land near the lake and the surrounding drier fixed dunes are grazed by cattle and horses in winter, with greatly reduced grazing in summer in order to improve the diversity of plants and to control more aggressive grasses. The tiny marsh snail, *Vertigo geyerii*, which is rare in Europe, occurs in abundance in the marshy area. The dunes provide valuable invertebrate food for foraging choughs, which are common on Donegal coasts. In summer, I walked along the sandy plain behind the dunes, where the short sward is ideal for swathes of colourful flowers like seaside pansies and bird's-foot trefoil. The current conservation management challenges at Sheskinmore and how they are being addressed were outlined to me by the Conservation Ranger, Emer Magee. She comes from a large family and she traces her interest in wild nature back to her childhood. 'My parents managed to take us on holidays to the most remote parts of the west of Ireland every year and, combined with frequent stays with farming relatives, this introduced me to wild places and farming before I had left primary school.' Emer explains that 'extensive cattle grazing is the optimal management for the flora and fauna around Sheskinmore. The animals are not housed and receive very little supplementary feed. They are taken off the marsh and dunes in the summer and put in the wet grassland areas, under an annual grazing agreement with seven local farmers.'

There are management problems at Sheskinmore that need to be addressed in an effort to find a balance. The fluctuations in the rabbit population over the last decade have had a detrimental effect on the dune habitats. The burrows are causing erosion and the disturbed sandy soil is being colonised by ragwort and burnet rose. The rabbits are grazing all year round, affecting the flowering plants and the grassland that is leased for winter grazing. The burrows are a threat to wintering cattle and horses that could be hurt by falling into them. The rabbit population is elevating the fox population, which also impacts ground-nesting birds. Various attempts by the Killybegs Gun Club and the NPWS trapper to shoot the rabbits have taken thousands of these mammals out of the site, but it has not had a significant impact on the population as far as I could see when I visited the reserve. The site is too big and the rabbit population too widespread to be able to control in a sustainable manner. Population changes are driven more by pathogens such as rabbit haemorrhagic disease and myxomatosis.

Greenland white-fronted geese were once regular winter visitors to the lake, but their numbers have fluctuated internationally since the 1980s, and now they are rare here. These days they are returning from the arctic with fewer juveniles each year, which has resulted in recent declines in overall numbers. When they do return, they prefer the re-seeded grass fields around Lough Swilly instead of Sheskinmore Lough and their traditional bog feeding sites. Here NPWS has worked with the local farmer Seamus Meehan to after-graze his cattle on the fields adjacent to the lough in September, which makes it more attractive for geese. Then in October a temporary electric fence is erected to keep the cattle out and to save that area for the returning geese.

Breeding wader numbers have also dropped significantly since the 1980s. This decline is a reflection of what has happened nationwide, as numbers of breeding lapwing here had declined to three pairs in 2009. In 2013, BirdWatch Ireland secured EU funding to erect a predator-proof fence around the main lapwing breeding area in the marsh west of Sheskinmore Lough. The organisation owns fifteen hectares of land here, and worked with NPWS and the other landowners to establish the project. Predation has been identified as a main cause for decline in breeding waders, and the electric fence is designed to exclude mammalian predators such as foxes and American mink. The lapwing numbers have increased to twelve pairs since the fence was erected, but predation continues to be an ongoing threat to the small numbers of waders breeding here.

Emer works closely with University College London (UCL) on research projects. Undergraduate and post-graduate students have carried out research based at the Field Studies centre at Sheskinmore. A doctorate has been carried out on the dune slacks, another on the hydrology of the lough. The Environmental Protection Agency (EPA) has funded research on the impact of grazing on the saltmarsh. In 2020, eight landowners around Sheskinmore Lough were approved for NPWS farm plans. From 2021, over 400 hectares have been farmed exclusively for nature as part of a five-year plan. Funds have also been made available to develop a management plan for the area. NPWS-led research has begun on the hydrogeology of the area, and this will complement the ongoing research by scientists from UCL.

Emer has been a Conservation Ranger in south-west Donegal since 2001, managing Sheskinmore among her many duties. On a personal level she finds it very challenging, but also rewarding, to manage the area for nature conservation. She says, 'I spend a lot of time alone here but I also enjoy the interaction with all the local people. I know the farmers and their families, the surrounding community, seasonal visitors, visiting researchers and the local historians. I am motivated by the desire to do some good and make a positive difference. I still want to learn so much and there is always so much to do. The more effort you put into trying to do something the more you want it to succeed.'

Gweebarra Bay

In a westerly gale I sat on the top of Dawros Head and watched the Atlantic breakers rolling in to crash against the rocks below. Suddenly an otter appeared out of the water and scrambled up the rocky face to take refuge from the storm in a reed-filled lake on the top of the headland. Walking further around the north side of the headland I came across a brood of kestrels with four well-grown chicks sitting in a row on a heathery ledge. My destination was the village of Portnoo which lies on a north-facing slope leading down to Gweebarra Bay. Praeger described this area 'as exemplifying what the visitor will find in this region':

> Below the scattered line of houses of Portnoo, green fields, white with pignut instead of daisies, slope steeply to a rocky shore, but to the right this gives way to a great beach of yellow sand, which at low tide connects with Inishkeel, where you will find the relics of a bygone ecclesiastical settlement with two churches and some incised crosses of primitive type .... Above the grassy slope of Portnoo, the scene changes abruptly, for everywhere there is rock and heather and lakelet,

> stretching southward to Loughros More Bay. Eastward stretches away to Glenties; westward it ends close by in a wild line of great cliffs which decrease as one goes towards Dawros Head. The heathland is dotted with granite boulders instead of bushes, and juniper replaces the familiar gorse. (WW)

Walking out across Narin Strand to reach Inishkeel, I had to check the tide tables carefully as the sea advances rapidly here across the level sands, and more than once I have been caught wading back to the mainland. At low tide, the northern side of the island has some wonderful rock pools filled with marine life, including the distinctive purple sea urchin. Offshore there are dense beds of kelp, the wide fronds waving underwater in the currents.

From here I walked to the east along wide sandy beaches and over impressive sand dunes near Clooney. This is the only place in Ireland where I have been challenged by a landowner claiming that he owned the beach because the sea had eroded his property and his land registry map now included the foreshore. In law, however, the area between high and low water mark in the Republic of Ireland belongs to the state and is therefore open to anyone to traverse.

Further east the sandy peninsula of Roshin Point is barely joined to the mainland by a narrow ridge of sand with the tide rising on either side. I sat on a high sand dune at the end of the point for a full tidal cycle to watch the movement of a group of harbour seals which regularly feed here and haul out on the offshore rocks at low tide. The fast-flowing waters at the mouth of the Gweebarra River are probably a good place for the seals to catch salmon and other fish which are concentrated in narrow channels. North of the river is Lettermacaward, where, on the flat fields behind Dooey Beach, I listened one summer in the 1980s to numerous corncrakes broadcasting their curious repetitive calls into the evening air. These birds have since vanished from many of their former haunts and are now largely confined to the north Donegal coast.

Arranmore

Arranmore, the largest of Donegal's offshore islands, is neatly framed in the view from Burtonport, where I caught the ferry on a summer's day. In the bay to the south several smaller uninhabited islands lie, including the flat sandy plain of Inishfree. This reminded me of hippy days in the 1970s when a commune set up home in a house in Burtonport. Up to thirty people lived in the commune in 'Atlantis House', which was brightly decorated, with eyes around the windows, astrological signs and yin-yang symbols on the walls. The group was known locally as 'the Screamers'. Primal therapy, including screaming, crying and unrestrained laughter were part of their lifestyle, as a way of expressing emotional needs, including anger and frustration. They were the first to highlight the problems associated with uranium prospecting in Donegal, but when other local people got the message and took up the challenge, the Screamers wisely stood back, as they knew their involvement would put many people off. The group's way of life raised objections locally, so in 1980 the commune relocated to several deserted cottages on the island of Inishfree. This aroused much curiosity in the national media, and the unwanted attention forced the commune to emigrate in 1989.

I was heading to Arranmore to meet a team of ornithologists from BirdWatch Ireland, who were studying the local chough population. These red-legged crows are plentiful on Donegal coasts because of the abundance of well-grazed maritime grassland. I walked with the birdwatchers to the lighthouse on the storm-lashed western side of the island. We found the aerobatic birds moving around the clifftops in family parties of four to six, the juveniles following their parents to the best foraging areas.

Walking the roads across the island, we passed groups of houses, some deserted and others only occupied in summer. In 1911, the population of the island had reached over 1,500 inhabitants who were mostly dependent on seasonal fishing and small-scale farming. The density of people in those days was quite evident from the extensive 'lazy beds' in small fields which were cultivated for potatoes and oats in the past but were now grassed over everywhere. By 1925, it was recorded that the fishing industry, based on a previously abundant herring migration, had now largely failed, and emigration to America or Scotland was the only prospect. Over the years, seasonal movements of farm workers to Scotland to work on the harvest had been a way of life for the people here and, even today, there are many families with Scottish relatives. In 1935, a disaster hit the island population when a boat carrying islanders returning to the harvest in Scotland set out from Burtonport in darkness. Hitting a rock, the boat sank, leaving nineteen passengers and crew dead and only one survivor.[73]

On the return trip by ferry I could see haul-outs of seals on various rocks, and I remembered a boat-based survey in the 1970s, when we found a good number of these animals here. Praeger recalled an incident where a hunter arrived on Arranmore to kill some grey seals that were raiding salmon nets and he confessed that 'a shrinking from the taking of life in any of its myriad manifestations has grown on me with the passage of time':

> I was glad to have no part in this, for no one who, like myself, has watched these beautiful animals on the coasts of half the counties of Ireland, could wish them ill; but then I am not dependent on the capture of salmon for my daily bread. I remember a deep clear calm gully, cliff-encircled; and in it a seven-foot seal at play – now resting on the surface, now diving to the bottom and groping fish-like among the long sea-weeds, clearly visible since the opposite cliff cut off the sky's reflection – a very beautiful sight: and I

> thought of this when I saw a full-grown seal and a very young one laid out on the boat slip on Arranmore.(WW)

Praeger was clearly shocked by the sight of the dead seals, but he knew that his view would not be shared by the west coast fishermen. The novelist Monk Gibbon also described a seal-hunting expedition on Arranmore in the early 20th century. The local men clambered into a cave where seal pups were lying and discharged their shotguns in the darkness. He wrote, 'It seemed inconceivable then that an ounce of lead, the dullest and heaviest, the most debased of all metals, should be capable of destroying the most delicate and beautiful mechanism in nature.'[74] Up to the 1970s seals in Ireland had no legal protection and bounties were even offered by the government for their heads. Today they are protected by Irish and European law, but licences are still issued occasionally for controlling individual seals at fishing nets, and there are frequent calls from the fishing community for larger culls.

North of Arranmore the coast breaks into a series of small islands off the Gaeltacht area of *Gaoth Dobhair*. Most of these held a small human population up to the early 20th century. The island of Gola attracted me because it had been deserted more recently. As I walked across the island from east to west, I passed many farmhouses that were still roofed and farm implements abandoned in field corners. I peeped through the window of a ruined schoolhouse, situated west of *Port an Churraigh*, that still held the wooden desks where up to seventy children sat in the 1940s. Stormy weather in recent years means the sea now comes right up to the schoolhouse door at high tide. It is weather-beaten, the roof has collapsed and it may be completely washed away in the coming years. The population of Gola rose steeply to 169 in the late 19th century, in response to a boom in the fishing industry, but collapsed again after that, and the absence of electricity was one of the factors that caused its final desertion in the 1960s. Today, there is a small but mainly seasonal population again.

Leaving the West Coast

Facing directly into the power of the Atlantic Ocean, the west coast is the most rugged and indented part of the Irish shoreline. It is also the longest section of shoreline and contains the greatest proportion of rocky cliffs, some of them among the highest and most spectacular in Europe. The west coast has about 80 per cent of all the larger offshore islands in Ireland too. Sand dune systems on the west coast are naturally more dynamic and mobile than those in the rest of Ireland due to frequent strong gales and storms. But the west has its quiet moods as well. I frequently find that I am the only person walking on a fine sandy beach in the middle of summer, as many of them are remote from roads. I love to walk around a rocky headland in west Donegal, where there are many secret coves, some known only to local fishermen and sheep farmers. This is a great place to find coconuts and other large seeds that have drifted across the ocean from the Americas. These strange drift seeds were often given mystical powers by local people, who had no previous experience of them.[75] The west coast is also frequently the first landfall in Europe for vagrant birds blown off course by Atlantic storms. In a warm summer of the mid-1980s I spent three weeks walking the sandy shores and grasslands of over fifty west coast bays from Malin Head in Donegal to the Aran Islands in Galway Bay. With some friends, I was looking for the small wading birds that make their nests in these attractive locations and whose young help to swell the flocks of autumn migrants moving south. Since those times, breeding wader populations on the unique sand dune machair of the west coasts have virtually collapsed. This is partly as a result of reduction in their range due to climate change but also due to a rise in predation and to more grazing pressure in the last decade, with sheep densities known to have increased markedly here. Total breeding wader populations on the west coast have declined by 62 per cent since 2009. Dunlin show the largest decline in numbers at 91 per cent. Oystercatcher, ringed plover, redshank and common sandpiper declined by over two-thirds, while snipe and lapwing both declined by between a quarter and one-third since 2009.[76] Other coastal sand dunes have been highly modified over the last few decades for conversion into golf courses, caravan parks and other developments, leaving less and less space for the native wild plants and animals.

'To hell or Connaught' was the cry of the Cromwellian invaders as they forced the native Irish from the good lands further east and confined them to the poor, mountainy land of the west coast and islands. The human population of the west coast was among the worst hit by the famines of the mid-19th century and there are many reminders in the deserted villages, small *clachans* and isolated farmsteads. The cultivation ridges known as 'lazy beds' are still evident in the grass and heather on many coastal fields, a reminder of the large farming population that once eked out a living in the remoter parts.

Today, the Wild Atlantic Way is known around the world as one of the most attractive long-distance tourist routes. As a marketing brand it remains highly successful but, in truth, the west coast has always been 'wild', and this should be valued for its own sake rather than just for its tourism potential.

North Coast

The completion and publication of the Clare Island Survey in 1911 marked a significant change of direction in Praeger's life and indeed for natural history in Ireland. Some of the leading Irish naturalists of the time – Richard Barrington, Richard Ussher and Robert Warren – died during this period, and the Field Club movement of the previous century had declined significantly. Not long afterwards Europe was plunged into a cataclysmic world war, while Ireland was engulfed by widespread political upheaval. With a German wife and a Teutonic-sounding name, Praeger may have felt the need to keep a lower profile. About this time, he turned to the less prominent pursuits of gardening and taxonomic botany, which led to the publication of two major accounts of plant groups. The reduced activity of the Dublin Naturalists' Field Club during the period of the 1914–18 war and the struggle for Irish independence was not mirrored in Ulster, and Praeger's attention again turned to his birthplace in the north, when he was elected Vice-President and subsequently President of the Belfast Naturalists' Field Club. He returned to lead field outings to Derry and Sligo, and frequently attended indoor meetings in Belfast despite continuing to live in Dublin.

From the 1920s, Praeger, now in his sixties and retired from his position in the National Library, resumed his passion for plant distribution. The treaty of 1921, resulting in the partition of Ireland, had dramatically changed the political landscape and led to the Civil War of 1922–23. This curtailed fieldwork and Praeger devoted more time to writing up the results of his extensive surveys. He rarely referred to the political situation in his writings and continued to regard the island of Ireland as a single unit for the study of natural history. By now, however, the natural sciences were evolving towards modern concepts of ecology, and the old amateur pursuits of merely collecting and listing of species were considered to be quite limited in value. So Praeger moved more in the direction of plant taxonomy, the study of the relationships between species.

In 1934, Praeger published another magnus opus, *The Botanist in Ireland*.[2] Press opinions were universally positive. *The Times Literary Supplement*, for instance, reported, 'Dr Praeger, the chief living authority on the Irish Flora approaches his subject place by place rather than plant by plant.' The *Irish Times* review said, 'Dr Praeger's new book is in every sense a masterpiece of botanical work. With the full knowledge of one who has tramped nearly every hillside and stream-side of the four provinces, he passes over no district without giving us a good bird's-eye view of its features as well as a trustworthy list of its most striking plants.'[3] The text followed a winding path through the island and in this respect it formed a scientific background to his most popular book, *The Way that I Went*, published just three years later. In the former he acknowledged the new scientific trends of the time, in that 'the claims of ecology as contrasted with purely floristic work have been recognised and an endeavour has been made all through to correlate vegetation with geography and geology'. In this context he commented that 'within the fretted coast-line of Ireland almost every sedimentary rock which has gone to building up the crust of the Earth is represented ... as well as many of igneous origin'.

Nor did he neglect the public interest in landscape, archaeology, history and folklore. In 1941, aged 76, he published a further book of reminiscences called *A Populous Solitude*. While this was still based on his comprehensive knowledge of the country, he also included stories gathered from people he met during his travels, some of them imaginary, such as a chapter on fairies at Hilltown, County Down. 'It was at this Hilltown ring-fort that one of Paddy Fitzpatrick's cows was shot by the fairies. I never got the details of this unlucky occurrence, which was not infrequent in the old days, when the magic cure – a flint arrowhead or elf-stone – would often be produced in corroboration of the story told.'(PS)

His chapter on the coast of Antrim reads more like a tourist gazetteer, with his impressions of the landscape of this popular route. Of the north Antrim coast he writes, 'you should not on any account miss the walk along these headlands, for you will not find anything better in the British Isles'.(PS) The western edge of this long north coast begins in north Donegal at Bloody

Foreland, from where it is possible to see the island of Tory facing directly into the wild Atlantic Ocean.

Tory Island

I strapped on my safety harness as the helicopter warmed up its engines in a car park on the edge of the coastal town of Falcarragh. I had been invited by the Tory Island Cooperative to visit the site for a new recycling facility, to assess if this would have any environmental impacts. While I had been several times before to the island, fourteen kilometres from the coast of north Donegal, the previous visits were always on a fast-moving ferry across a rough sea. This was an unmissable opportunity to see a spectacular part of the coast from the air.

With the dramatic outline of Tory in the distance, we flew over three other smaller islands – Inishbofin, Inishbeg and Inishdooey – on the way. Exposed to everything that the Atlantic Ocean can throw at it, Tory has some spectacular rock structures on the north and west sides of the island. The hard quartzite rocks have been eroded to produce offshore stacks, arches, caves and cliffs, some of them reaching to almost 100 metres in height. The most impressive is the knife-edged promontory of Tormore, which reaches out to the north-east of the island. Praeger wrote, 'it is strange to find so desolate and tempestuous a place has been long inhabited and even fortified'.(WW) From the clifftops the land slopes away to the south-west like a listing raft in a vast ocean. On its surface I could see two clusters of houses known simply as West Town and East Town. An early Christian resident of the island was St Colmcille, who founded a monastery here in the 6th century. The most obvious remnants of this are the round tower and the distinctive T-shaped cross which face visitors as they step off the boat at the pier in West Town.

Until the 19th century, the people of the islands lived mainly by farming but, when the Congested Districts Board took over control of the island, they built harbour facilities and provided grants for the purchase of boats and nets. This led to a greater commitment to fishing in the island's economy and less intensive farming practices in response. Tory farmers used to have a version of the ancient open-field 'rundale' system of allocating plots for cultivation annually on a community basis. But widespread cultivation of the land for potatoes and oats was largely changed to extensive grazing. By the time of my visit, farming had been all but abandoned to rushy grassland, and I only saw a single small flock of sheep.

Low-intensity farming has led to the islands off the north and west Donegal coast now being among the few places in the country where corncrakes return to breed each year. Here the numbers are still impressive, with some forty-three calling birds recorded on Gola, Tory and Inishbofin islands in 2017. A scheme run by the National Parks and Wildlife Service (NPWS) aims to protect all suitable corncrake habitat within a 250-metre radius of any male that calls over five consecutive days. This is generally accepted as the area within which the female may be nesting. Research shows that home ranges of each pair are very variable, from less than one to over fifteen hectares. The scheme promotes the development or retention of early and late cover, as the vegetation in the meadows is generally not tall enough to provide cover for corncrakes when they arrive in early summer. Similarly, after meadows are cut, these provide cover late in the breeding season.

Marie Duffy, who managed the project for the NPWS, says, 'we work to enhance and create areas of tall cover and encourage farmers to delay mowing.

The work on the islands is all encompassing, creating suitable corncrake habitat from abandoned land. We are trialling new tall plant species here such as reed canary grass, iris, nettles and artichokes, to see what grows well in such an exposed habitat and what is suitable for corncrake.' The management work goes on right though the year – clearing vegetation, fencing and creating nettle patches (winter); ploughing, sowing and planting other early cover species (spring). Crops grown include a black oat mix, barley/oats with kale and triticale, potatoes and a grass mix favouring tall coarse grasses. Some areas within the fenced perimeter are left fallow after vegetation clearance, as they are being managed for breeding waders.

On my visit to Tory Island I met up with Marie to see what attracts the corncrakes to this remote spot off the north coast of Donegal. In her words, 'the islands have become the focus for conservation of the corncrake in Ireland as the farmers here have largely stopped cutting grass because there are few overwintering livestock that require fodder. The problems are now lack of early and late cover and the overgrazing of some meadows in summer. We are working to try and restore some of the habitat that the birds need.' As one of the few globally threatened species in Ireland, this is a high priority for conservation.

By 2018 it appeared that the scheme was having some positive results, with the number of calling males in the country increasing by 8 per cent to 151. In 2019 this number increased to 162, but fell again to 146 in 2020. However, this represented an overall 20 per cent decline on the 1993 total, when conservation measures were first introduced in Ireland. Since 2012, the State has spent several million euro on this scheme, with hundreds of farmers in the west and midlands availing of grants to adjust the management of grasslands for the birds. The most recent increase in numbers was attributed to the exceptionally warm weather that summer, and was the first rise in numbers since 2014. The Corncrake Project Annual Report for 2018 confirmed that Donegal remains the national stronghold for the bird, where a total of ninety calling males were recorded that year – fifty-nine of these on the county's offshore islands. The recently launched Corncrake LIFE project aims to deliver actions for this threatened species across a total of 1,000ha in eight areas down the west coast. This is to be achieved through a combination of habitat management agreements, land leasing and strategic land purchase. This scheme is like the emergency ward of a hospital where the patient is on life-support. The type of farming that used to support the corncrake has largely disappeared from Ireland, and only specially targeted measures will prevent the species becoming extinct in the few areas where it still holds on.

Bloody Foreland to Fanad

Bloody Foreland forms the very north-west corner of Donegal, where the wild Atlantic waves have built up a massive storm beach of large boulders. Its name is thought to relate to the red sunsets so often seen out over the Atlantic rather than any bloody battle or murder. Following the coast a little further to the east, I reached the huge mobile dune system that lies off the townland of Gortahork. I remember walking the coast here in the 1970s when a local farmer was cutting marram grass from the dunes and using it to thatch his farm buildings. The tough dune grass was once widely used for thatching on the west coast but this tradition has now died out.

The next major headland on the north coast of Donegal is Horn Head, which stands up to 250 metres above sea level. Praeger wrote:

> It is the massive architecture of Horn Head that both produces so imposing an effect, and at the same time makes it difficult to realize the scale on which it is built – a similar impression is sometimes produced by a well-proportioned building of great size. The quartzite forms solid bastions rising from base to summit, separated by deep vertical rifts, the whole being tilted over at an angle. The ledges are a breeding-place for tens of thousands of sea-fowl and as one passes in a boat the air is thick with razorbills, guillemots, puffins, gulls of several kinds and other species in smaller numbers.(WW)

I walked along the clifftop here in autumn and watched as exhausted redwings, each slightly smaller than our native song thrush, flew in from the ocean and landed in the heather. They had migrated non-stop from Iceland ahead of a looming arctic winter. Linking Horn Head to the mainland at Dunfanaghy is an extensive area of sand dunes, which was once so mobile that it swamped houses and farms, both here and on the opposite side of Sheep Haven. Praeger knew the damage that had occurred and he recorded:

> About Dunfanaghy blowing sand has long proved a serious menace, as it has done also at Rosapenna. At the latter place, in 1784, it finally overwhelmed Lord Boyne's house, demesne and garden, as well as sixteen farms, burying some of them to a depth of twenty feet. This catastrophe is stated to have been due to the killing of the foxes, which preyed on the rabbits inhabiting the dunes. The rabbits increased greatly in numbers, and burrowed and destroyed the turf to such an extent that, during a dry stormy winter, the dunes were themselves destroyed, the material blown inland and deposited to the amount of thousands of tons on the demesne and farmland.(WW)

To the east the coastline is deeply indented with three large inlets: Sheep Haven, Mulroy Bay and Lough Swilly. Just behind the little pier at Portnablagh is a classic example of a blowhole where the roof of a cave has collapsed and I could stand by the hole listening to the waves roaring among the rocks below.

I felt a great sense of excitement when entering Mulroy Bay by boat for the first time. Praeger referred to it as 'that extensive land-locked, most complicated piece of water'.(WW) The bay has a very narrow entrance and a long contorted shape with many small islands and much unspoilt landscape on its shores. The channel varies in width and depth, with some narrow points only 100–150 metres across, where the current is very strong. It stretches south for approximately twelve kilometres ending at the village of Milford. This was the location of a large flour mill and bakery up to the 1970s, but I can also remember the discharge of enriched water that it produced in the bay. In 2010 a new high-level road bridge was opened across the northern part of the bay, replacing the old ferry that plied across the water up to the 1960s, connecting the communities of Carrigart and Fanad. At 340 metres in length, this is one of the longest road bridges in Ireland. This reduced the driving time around the bay from up to two hours to less than ten minutes.

Due to the unusual tidal currents here, Mulroy Bay contains a range of different sediment types, which includes coarse sand, beds of calcareous algae called maërl and a variety of exposed and sheltered underwater reefs with strong to weak currents. The varied habitats support many marine communities, with high species diversity. Rare marine species found in Mulroy Bay include a little fish – the Couch's goby – a file shell, an anthozoan, a hydroid and a red seaweed. In the more sheltered areas with less current, dense kelp forests occur, and these have a greater variety of sponges and solitary sea squirts. Otters are sometimes seen here, and I remember finding a group of harbour seals hauled out on various islands in sheltered parts of the bay. There were once breeding terns on the islands but predation by introduced mink and disturbance from aquaculture caused them to desert. Feral greylag geese are now breeding in some numbers, and it is not unusual to see a flock of over 100 birds at the end of the breeding season. A large population of native scallops is also present here, and this is now commercially exploited. Despite being protected under the EU Habitats Directive the NPWS has acknowledged that aquaculture, scallop dredging and seaweed harvesting here pose a threat to the ecological value of the area.

Lough Swilly

It was in Lough Swilly, at the small pier of Rathmullan,

that I joined an international group of biologists who set out to find and count harbour seals in the more sheltered waters of north Donegal. We launched two powerful inflatable boats from the slip and scoured the coastline looking for the haul-outs of these marine mammals. Just offshore a tiny storm petrel danced about on the waves. From its opening to the sea between Fanad Head and Dunaff Head, Lough Swilly stretches inland some forty kilometres into the heart of Donegal, almost to the town of Letterkenny. The inner areas are shallow, sheltered, muddy estuaries where the rivers Leannan and Swilly discharge their loads of silt into the tide. The extensive mudflats are packed with marine worms and shellfish, attracting tens of thousands of waterbirds in winter, as it is one of the nearest landfalls in Europe for birds migrating south from Iceland and Greenland.

It was also here at Rathmullan that a watershed event in Irish history took place. Up to the end of the 16th century, much of the province of Ulster was ruled by Irish chieftains but, following their defeat by English forces in 1603, Rory O'Donnell, the earl of Tyrconnell and Hugh O'Neill, earl of Tyrone, with some of the leading Gaelic families in Ulster, boarded a French ship and travelled down Lough Swilly into exile in mainland Europe, where they hoped to raise Spanish support for their cause. The so-called 'Flight of the Earls' marked the end of the old Gaelic order and the beginning of British colonisation of Ulster, which was to have major implications for the whole island, not least in the partition of Ireland in 1921 and the Troubles of the late 20th century. Walking along the beach at Rathmullan I tried to imagine a tall ship sailing hurriedly down the lough with English soldiers being marched in on all sides to garrison the towns. From then on, through the Napoleonic wars of the early 19th century, Lough Swilly was to become an important military base, with extensive fortifications. During both world wars the lough took on a strategic role as a sheltered base for Royal Navy warships defending the American merchant shipping passing Ireland to the important ports of Liverpool and the Clyde.

In the 19th century low-lying farmland on the western side of the lough was created by draining and infilling some arms of the estuary, but the sea walls north of Ramelton have since broken down and the sea has reclaimed a large area, turning it back into mudflat and saltmarsh. With modern sea level rise this process is likely to be repeated in many places around the Irish coast, as the value of coastal farmland is often less than the cost of repairing the sea defences. About halfway along the eastern shore the large island of Inch was linked to the mainland by two causeways, and the area between was partly pumped out and claimed for agriculture. This area, known as Inch Levels, is now one of the best sites in north-west Ireland for wintering swans and geese. As the autumn days close in great V-shaped skeins of whooper swans and greylag geese jet in from their Icelandic breeding grounds. Here they are attracted to the rich feeding that they find on these large, flat fields of grass and cereals. Walking along the raised embankments I could hear the swans trumpeting out their evocative calls, reminding me of the sounds of lonely areas of tundra in Iceland during the summer.

Inch Island was also the location in the 18th century for a commercial fishery station established by one of the leading merchants of the nearby city of Derry. A large salting station was established here to cure up to 100,000 herring per day, with forty women and children employed to gut the fish and five coopers on hand to make the barrels in which the fish would be preserved. In 1776 some 1,750 barrels of salted herrings were exported to the Caribbean, where they would have been used to feed the burgeoning population, including slaves working in the plantations. At this time there were up to a thousand fishing boats in use in the lough, which gives some idea of the size of the herring shoals which some fishermen said were difficult to row through. By the early 19th century the herring shoals had dramatically declined, in an early example of the impact of overfishing.[4]

Overfishing is still a problem in Lough Swilly. This is one of a decreasing number of bays in the west of Ireland where the native oyster still occurs. This shellfish was once widespread in the country and, growing in abundance in shallow water, it was easily collected until it became extinct in many areas. Lough Swilly is now a protected area, and in 2013 a five-year plan for conservation of the native oyster was prepared, but so far this has not been implemented. The current unlimited fishery has significantly reduced native oyster numbers, and there is now a real risk of population collapse. In addition to the fishery, the culture of non-native Pacific oysters is increasing in Lough Swilly, and this is impacting on the native oyster population due to escapes from shellfish farms.[5] It is now known that feral Pacific oysters are well established in the Shannon Estuary, Galway Bay and Lough Foyle, as well as Lough Swilly.[6]

The eastern limit of Lough Swilly is formed by Dunaff Head, which holds several rare northern plant species, including roseroot and rock samphire. This made it a happy hunting ground for naturalists like Praeger and Henry Hart, author of *Flora of the County Donegal*. In 1885, Hart, botanising on Slieve League at the south-west extremity of the county, met an old man and a small boy carrying armfuls of samphire up a steep path, eating as they climbed. This species was widely collected and sold as food or for herbal remedies.[7]

Inishowen and Inishtrahull

Between the long inlets of Lough Swilly and Lough Foyle is the Inishowen Peninsula, where the people have many of the characteristics of an island community. I felt a sense of satisfaction to stand on the most northerly part of the Irish mainland beside the lighthouse at Malin Head. From here I could see the outline of Tory Island to the west and Islay fading away into the Scottish mist to the north-east. Below me was one of the finest examples in the country of a raised beach, with its fossilised cliff-line well above the present sea level. This illustrates dramatically how the northern coast of Ireland is still 'rebounding' more than 15,000 years after the last Ice Age, when it was depressed by the weight of ice, several kilometres thick. I liken this to a mattress springing back to shape after a sleeping person has risen in the morning.

Inishowen has some fine walking country. Praeger criss-crossed the mountains in the centre, including Slieve Snacht, and scoured the area for rare plants. He advised visitors to 'walk north from Malin village along the west side of Malin Head (where you will sleep) and back along the cliffs to Glengad, and again you will have seen a splendid stretch of precipitous coast'. He commented that 'the Inishowen flora is, as one might expect, northern rather than southern, and boreal plants such as the Scottish lovage and the sea gromwell or oyster-plant are frequent on its exposed shores'.(WW)

I had to climb higher till I could clearly see the island of Inishtrahull with the most northerly lighthouse in Ireland. These waters around the northern point of Ireland are a place where strong tides rip past the headland from the Atlantic Ocean and upwelling from the sea floor carries a rich harvest of plankton, the microscopic life that forms the foundation of all marine biodiversity. Here too is a wonderful place to see basking sharks, the largest fish in the Atlantic, when they congregate in feeding shoals. To have a closer look myself I went to the pier at Culdaff, where I was told I would find a boat to take me to the island. As it happened that day, the ferry was also being used by a team from the Commissioners of Irish Lights on one of their regular servicing trips to the automatic lighthouse. Landing on a high pier in a sheltered cove on the north-east side of the island, I walked alone around the cliffs and imagined life in this remote spot before the days of modern communications. There was once a small community of islanders living here, and the ruins of some of their houses and abandoned farmland remain to this day in the low-lying centre between two hills. There was a large increase in the island population in the 1880s, with as many as thirteen families resident on the island up to the 1920s.[8] At the eastern end there is a small graveyard, and the inscriptions, though now heavily weathered, tell a tale of infant mortality that was common in remote areas in previous centuries. In the 1890s the Congested Districts Board set up a fish-curing station on Inishtrahull to give some employment in this remote location.[9]

The island is made of a hard rock called gneiss, of the oldest type found in Ireland. The first lighthouse on Inishtrahull was built at the eastern end of the island in 1813 at the behest of the Royal Navy, which was then using the nearby Lough Foyle as a base for its North Atlantic operations. The last of the islanders moved to the mainland in 1928 following a tragedy when a mailboat from Malin was swamped and a young islander drowned. This left the lighthouse keepers as the only occupants of the island. In the 1950s it was decided to reposition the light and a new tower was built on the west end of the island, making it one of the more modern lighthouses in Ireland.[10] When the lighthouse was automated in 1987, the island was left to the birds, which now include a sizeable population of eider ducks in summer and flocks of barnacle geese in winter.

After a while I sat looking out through my binoculars over a calm summer sea. In the distance I could see some slow-moving fins, but it was difficult to distinguish the number of basking sharks here, as both the dorsal fin and tail flukes are often visible above the water. I could imagine that below the surface there was a gathering of these enormous fish circling slowly with open mouths the size of shed doors. Although they have several rows of small teeth, basking sharks feed mainly with the use of gill rakers, through which they filter plankton gathered in through the open mouth like a giant vacuum cleaner. The primary food of the basking sharks is zooplankton, specifically copepods, which are tiny crustacea, and current research shows that the presence of one particular copepod species often determines where the fish feed. Emmet Johnston of the NPWS has estimated that up to a quarter of the world's basking shark population moves through the sea off the north coast of Ireland at certain times of year. The average length of adult basking sharks is five to seven metres, although some specimens have been recorded at eleven metres, the size of an ocean-going yacht. The liver, for which these fish were originally hunted, is estimated to make up approximately a quarter of the body weight. This organ acts as a float to keep the shark at near neutral buoyancy. The basking shark uses its large tail to power through the water with a slow, sinuous movement but the fins on its back and tail are the only visible parts of the huge body.

I called in to visit Greencastle, on the most easterly part of Inishowen, to see where an almost forgotten wooden traditional boat was built. The Drontheim (also known as a 'Greencastle yawl' or 'north coast yawl') is an open, double-ended and clinker-built boat, generally about eight metres in length, which can trace its origins back to the Viking boats of the 8th century. Its name is derived directly from the town of Trondheim in Norway, from where the original boats were imported. These boats were once a familiar sight on north coasts from Donegal Bay to Ballycastle in County Antrim. They were the traditional fishing sailboats for many generations of fishermen in small communities on the rough waters of this part of the Atlantic. With the arrival of marine engines, these sailboats disappeared virtually overnight, and many were left to rot on piers and in fields all along the coast. From the 1980s a restoration movement began, with new Drontheims being built from the old designs and some surviving original boats restored.[11]

Lough Foyle

From the harbour at Greencastle, I looked across the

narrow strait that marks the border with Northern Ireland. Born near Belfast and living in Dublin, Praeger was clearly very aware of the tensions between north and south. He rarely mentioned politics in his writings, but there is one exception towards the conclusion of *The Way that I Went*, published in 1937, just sixteen years after partition of the island:

> Ireland is a very lovely country. Indeed, there is only one thing wrong with it, and that is that the people that are in it have not the common-sense to live in peace with one another and with their neighbours. Past events and political theory are allowed to bulk much too large in our mental make-up, and the result is dissatisfaction, unrest, and occasionally shocking violence. That 'frontier' [the border between Ireland and Northern Ireland], which is a festering sore in Ireland's present economy, was the noble gift of an English Government – but Ireland brought it on herself.

I crossed into Northern Ireland and went to visit the northernmost part of County Derry at Magilligan

Point. I could have saved myself a lot of driving around Lough Foyle if I had taken the seasonal car ferry that runs from Greencastle to Magilligan Point. Londonderry is the official name for the county, as shown on Ordnance Survey maps, but the urban area is often called Derry/Londonderry or 'Stroke City'. To the south of where I stood a vast area of shallow water opened out into the sheltered waters of Lough Foyle. Praeger viewed the same area from the top of the nearby mountain of Benevenagh.

> From the summit of the hill we get a bird's-eye view of the remarkable flat sandy triangle, with five-mile sides, known as Magilligan, which almost blocks the entrance of Lough Foyle. If we descend the hill and walk over its sandy surface we can study the successive ridges, parallel with the outer shore, which show its growth mile after mile towards the open sea, nor does this process appear yet to have ceased. As to why it was so formed one may well puzzle; as to when, it is of course comparatively new – post-Glacial at least; but where did such a prodigious quantity of sand come from?(WW)

The question raised by Praeger has been partly answered by environmental scientists from the University of Ulster who have carried out extensive studies of dune systems all around the Irish coast over the past half-century. Bill Carter and Peter Wilson showed how beach ridges formed as the land rebounded after the weight of glacial ice was removed and the sea withdrew northwards sometime after 6,500 years ago. This created a huge area of flat beach from which wind-blown sand was later deposited on top of the ridges at least 1,790 years ago, forming the original sand dunes. This was followed by several more periods of active dune formation between 1,200 and 600 years ago, and again during the 17th and 18th centuries.[12] There is a Martello tower from Napoleonic times at the end of Magilligan Point. Because of the sedimentary cycles, it can be found beside the beach or up to 100 metres inland.

North of the city of Derry, on the eastern side of Lough Foyle, a railway embankment runs along the shoreline, cutting off large areas of reclaimed farmland from the tides. Beyond the sea wall huge sandbanks and mudflats stretch out towards the centre of the lough, and these are attractive to wintering flocks of brent geese that drop in here after their long migration from breeding grounds in arctic Canada. These small, dark-coloured geese form large family-based flocks and, as I walked along the old railway line, I could hear their 'conversations' carrying across the wet sand. I was making a visit to Lough Foyle to map the extensive beds of eelgrass, the main food of the geese. These marine grasses occur between the tides in areas well sheltered from tidal currents. Although modern methods, using satellite imagery and drone photography, are now available, this was in the early 1990s, when the only way to map the outline of the grass was on foot, walking across the soft sand and mud at low tide with a tape measure and compass. With my colleague Micheál Ó Briain, I mapped eelgrass beds in dozens of estuaries all around the Irish coast, as well as reviewing existing information from other studies, both historical and modern.

North Derry

Situated on the north coast near Coleraine, County Derry, and nestled between the beaches of Benone and Downhill, is the sand dune system known as Umbra. I went there first in the 1970s, as it was one of the original nature reserves established by the conservation body Ulster Wildlife. A number of coastal habitats of special conservation value and many distinctive species make this a very important site. Approximately forty-five hectares in size, one of the chief attractions is the summer display of flowering plants in the meadows and sand dunes, and these in turn support an incredible diversity of insect life – butterflies, moths and bees being the easiest to observe. The nature reserve is noted for a number of rare plants: bee orchid, pyramidal orchid, fragrant orchid, frog orchid, marsh helleborine, adder's tongue, moonwort, hairy rock cress, downy oat-grass, fern grass and early forget-me-not. Interesting butterflies such as Real's wood white, grayling and dark green fritillary are found here, as well as the rare small eggar moth. It also has several of Northern Ireland's priority bird species – skylark, mistle thrush, song thrush, bullfinch and linnet.

Current conservation work by Ulster Wildlife at Umbra was outlined to me by Andy Crory, Nature Reserves Manager with this busy conservation organisation. Andy has a team of assistants supplemented by one-year interns from all over Europe. More than half a hectare of sea buckthorn was removed from the dunes in 2018, thanks to funding by the Landfill Communities Fund, with follow-up works planned for future years. This non-native species was a direct threat to the species-rich dune grassland here. The non-native pine plantation at Umbra has been removed too – a landscape-scale change to this site. The project was completed thanks to a partnership with Northern Ireland Environment Agency and Drenagh Sawmills. Several hectares of trees were removed, which will benefit both the grassland species and the wet dune slacks that make this site so important. Andy, who is an enthusiastic recorder of moths, has a vision of an open dune landscape here which will be maintained into the future with sensitive grazing. Ulster Wildlife has produced a management plan for the site which will make it easier for future staff of the organisation to continue the good work. Sand dunes are special ecosystems in the coastal environment, with a unique ability to restore themselves after erosion pressure, given space and freedom to move. Enlightened management is the key to protecting the few remaining dune systems, like this one, that remain undeveloped.

Viewed from the north coast, the towering

Benevenagh mountain marks the north-west corner of the basalt plateau which dominates the county of Antrim and stretches into Derry. On this mountain, Praeger searched for a number of rare alpine plants – mountain avens, purple saxifrage, cushion pink, twisted whitlow grass and others. Moving east, the coast road runs along the clifftop at Downhill, giving spectacular views back to Inishowen Head to the west and Benbane Head to the east. Perched on the cliff edge here is a strange circular building in classical style known as Mussenden Temple. Built by the Bishop of Derry, who owned the nearby Downhill Demesne, the design was based on the Temple of Vesta in Italy, and once held the Earl Bishop's library. Now protected by the National Trust, the inscription on the building reads 'Tis pleasant, safely to behold, from shore, the rolling ship and hear the tempest roar'.

To see if it was 'pleasant', I sat above the cliffs here in the setting sun and watched the gathering of huge flocks of starlings, coming from all directions between east, south and west. In the sky, they circled and massed together until they formed an enormous flock known as a murmuration. For a period, the flock wheeled overhead, constantly changing shape like a giant amoeba and then, in a few seconds, it disappeared beneath the cliffs into a cave where thousands of birds sat chattering along ledges to roost for the night.

I had cousins who lived near Castlerock and we played as children on the beach and sand dunes here. Praeger had similar memories from almost a century earlier:

> Castlerock, a couple of furlongs of scattered villas facing the sea, has changed but little in the last half-century. For this I am grateful, for it is bound up with many of my happiest childish recollections. Hither, away back in the seventies and eighties [1870s and 1880s] we used to have a family migration in August. Here we learnt to swim without fear in rough water; here we found hare-bells and gentians, and white and yellow water-lilies difficult to gather without a wetting; and the sand-dunes, now thronged with golfers, were lonely and exciting places, ideal for the hunting of Red Indians and for expeditions through untrodden mountain passes. And they yielded wild strawberries too, in profusion, a gift beyond price for starving explorers.(WW)

Today the beach at Castlerock is largely unchanged since Praeger's times, but the 'scattered villas' have morphed into a large group of modern bungalows, many built as second homes.

Bann Estuary

The River Bann, which rises in the Mourne Mountains at the southern margin of Ulster, finally reaches the sea here, having passed through Lough Neagh in between. The long estuary runs for six kilometres, entering the saltwater between long sea walls. Just upstream, on the edge of Coleraine, is Mount Sandel, the earliest known human settlement on the Irish coast, which was excavated by Peter Woodman in the 1970s, so Praeger was probably unaware of this important site. In the centuries following 8000 BC early people returned repeatedly to this site, occupying several phases of circular huts. Built of substantial posts and roofed with turfs, rushes or perhaps even sealskins, the huts had a central hearth, so the buildings must have been filled with smoke. Inside the house sites the archaeologists found hazelnut shells and burnt animal bones, while outside were several large pits used for rubbish disposal or food storage. Here too were abundant flint implements, including both finished and partly worked tools. These Stone Age people would have inserted the flint blades, axes, scrapers and other tools into handles made of wood or deer antler. The animal bones were dominated by wild pig, with some mountain hare and either dog or wolf. Some birds, including waders and wildfowl, were also hunted on the estuary, while fish remains found include salmon, trout and eels, many of which may have been caught in fish traps. Mount Sandel is generally thought to have been a kind of base-camp from which these early people would have ventured out on hunting and perhaps trading expeditions.[13]

The Bann Estuary includes Portstewart Dunes, which are managed by the National Trust. The dunes owe their existence to plants such as marram grass, which can grow in the harshest conditions, trapping blowing sand and forming embryo dunes. Further back from the sea the dunes become more stable with age, supporting different communities of plants, resulting in a site so rich in species that the Bann Estuary dunes have been recognised nationally and internationally for their importance. Finding a balance between the dunes becoming so stable that they lose their special character and so unstable that they are damaged and eroded is central to their ecological management. Historical management has included the planting of non-native sea buckthorn, which has spread across the dunes at the expense of the native flora. Human impacts have damaged the vegetation on the seaward face of the dunes, reducing the formation of new dunes and the protection from the waves of storms. The key focus of future management is mitigating past human activity, while allowing nature to thrive and be enjoyed. I finished the day sitting at the top of Portstewart Beach watching the sunset to the west and enjoying a feast of the best fish and chips in Northern Ireland from Harry's Shack.

As well as being an important navigational channel, the estuarine banks of the Bann have a range of important habitats, not least an extensive area of saltmarsh. Almost ten per cent of Northern Ireland's coastal saltmarsh is found on this site, and it is home to a wide range of specialist plants and animals perfectly adapted to living in an environment of regular inundation by salt water.

A short distance to the east of the Bann Estuary, at Portrush, there are basalts as well as shales and clays, from the Lias period, that contain numerous fossils. Here Praeger recalled 'one of the great scientific controversies of the past'. In 1775, a dogmatic geologist named Werner was appointed head of the School of Mines at Freyburg (now Freiburg), Germany. He did not believe that any rocks except recent lavas were of volcanic origin. He and his followers were called 'the Neptunists', as they believed that all basalts had been precipitated in water, and they were opposed by 'the Vulcanists', who regarded these as igneous rocks that had originated from ancient volcanoes. Praeger wrote:

> Questions of religious belief were by no means excluded, for those who would not admit that all rocks were sedimentary were accused of attempting to undermine Holy Writ, in that their tenets were incompatible with belief in the universality of the Flood. ... At Portrush an intrusion of molten lava ... so baked the Lias (clay) that it became a hard dark flinty rock, which was mistaken for one of the volcanic series: and the inclusion in it of numerous fossils was hailed with delight by the 'Neptunists' as proof that the basalt itself was of aqueous origin as contended by Werner and his followers.(WW)

It took until 1835 until there was general agreement about the volcanic origins of this rock, which, as Praeger wrote, 'still outcrops boldly on the beach at Portrush, a monument to a famous battle'.(WW)

Giant's Causeway

One of the geological wonders of Europe – now a World Heritage Site – is the Giant's Causeway, which lies about ten kilometres east of Portrush. The site was generally only known to local people until the late 17th century, when a number of prominent clerics, scholars and natural philosophers 'discovered' the unique rock formations. In 1739 Susanna Drury, a previously unknown Dublin artist, travelled to the Causeway to paint the landscapes, and some engravings of her work were circulated throughout Europe, arousing great interest among scientists. By the mid-1800s, with the proliferation of railways throughout the country, tourism became a popular activity for the wealthy, and the Giant's Causeway featured prominently in many guidebooks.[14]

The most famous image of the Causeway shows the regular hexagonal columns, in some cases up to twelve metres high, which outcrop in a number of cliffs.

The basalt, of which they are formed, was originally molten lava which was forced out onto the earth's surface. As the hot liquid cooled and hardened in the air, shrinkage caused cracks to appear and the regular geometrical shapes emerged, each column locked into those surrounding it like honeycomb in a beehive. I walked down the track that leads from the car park to follow the shoreline and a whole vista of different shapes and patterns spread out before me. The most popular features, known by romantic names such as the organ pipes and the chimneys, were crowded with tourists, all brandishing cameras, so I kept walking and soon found myself alone on this beautiful coastline. I had to agree with Dr Samuel Johnson who, when asked what he thought about the Giant's Causeway, proclaimed 'Worth seeing, yes; but not worth going to see'.

In the 19th century the coast at the Giant's Causeway would have been marked almost continuously by columns of white smoke from the fires used to burn seaweed collected from the shore. The rocky seabed here is covered with dense beds of large brown seaweeds, collectively known as kelp, which waves back and forward in the tides. Their anchorage or 'holdfasts' are sometimes ripped from the rocks by winter storms and great banks of tangled weed cast up on the shore. A century ago this seaweed was a valuable commodity which was collected by hand, carried in wicker creels made from hazel rods and laid out to dry on specially built stone walls. By the summer, the dried fronds were piled onto stone kilns and burned until the remaining residue, looking like a bluish sticky toffee, cooled and hardened. This product, also called kelp, was cut into blocks and sold, as it contained a concentrated source of sodium and potassium, chemicals used in bleaching and for making soap and glass. The value of the kelp was further enhanced by the discovery that it was a source of iodine, then widely used in medicine and early photography.[15]

The flora of the basalt cliffs of north Antrim is rich and varied as they are largely ungrazed. On the Shepherd's Path which leads down to the shoreline, the vegetation contains plants more typical of woodland mixed with the familiar maritime species. In early summer, I found primroses, bluebells and early purple orchids growing among sea pink and sea campion flowers. There are caves here, which sheltered some of the early people who lived on these coasts. Among the bones that archaeologists have found on cave floors are those of the great auk, a flightless seabird that is now extinct throughout its range. It's possible that it was hunted for food here on the north Antrim coast while nesting among the related razorbills and guillemots.[16]

The Causeway coast path is one of the most varied and interesting coastal walks in Ireland, passing by the spectacular Carrick-a-Rede suspension bridge, once used by salmon fishermen to reach their nets, and the dramatic ruins of Dunluce Castle which hang above the Atlantic waves. My longest day's walking was rewarded with a visit to the Bushmills Distillery. Founded in 1608, Bushmills is the oldest licensed distillery in the world. Here malted barley is distilled in the original copper stills to make a single malt that is world famous. In 1885, a disastrous fire destroyed most of the distillery, but it was rebuilt and now continues the tradition of earlier centuries.

Finally, I reached the town of Ballycastle, nestled into the northern edge of the Antrim Plateau. Just outside Ballycastle, I came across the North Star Dyke, evidence of violent seismic activity about sixty million years ago. The name arose because it points roughly towards the North Star and, at shorter range, aligns with Church Bay on Rathlin Island. It is a four-metre-thick olivine dolerite intrusion where the sandstone has been washed away from it on both sides, leaving the dyke resembling a stone pier. Dolerite is a volcanic rock, very similar to basalt, containing crystals indicating that it cooled a little more slowly. Ballycastle is the site of the annual Lammas Fair. Frequently given the adjective *Ould*, this is one of the oldest fairs in Ireland, where animals would have been bought and sold and competitions held each year. Here, at a street stall, I sampled local products called Dulse and Yellowman, an unlikely pairing of honeycomb toffee and dried seaweed.

Rathlin Island

At Ballycastle I took the regular ferry that crosses the narrow strait between the mainland and Rathlin Island. Praeger was particularly fond of Rathlin at the north-east corner of Antrim, where he spent many days of his youth exploring and learning to understand the natural world.

> These island crossings were for the greater part among the most exhilarating things in life. Who

> would not wish to find himself at dawn on a June morning in the rushing tide that eddies between Rathlin and Ballycastle, with Fair Head rising like a black wedge to the eastward, and the sunrise coming up over the Scottish islands?(BS)

In his old age, Praeger wrote about the coast of Antrim and vividly remembered standing on a clifftop where he rejoiced 'in the boisterous sparkling northern sea, the imposing precipitous headlands one beyond another'. Following a colourful description of the basalt, his eyes lingered 'longingly over Rathlin, set in the blue sea'. He then recalled a remarkable experience on the island many years earlier:

> Here with two other youthful adventurers, I experienced for the first time, in an open boat, a really bad sea which made us, and our hardened boatman, very glad indeed when we reached Church Bay. Here we saw for the first time vast colonies of breeding sea-birds tight-packed on stack and precipice. And here I had my earliest experience of rock work. It was 'Paddy the Cliff-climber' who helped at that initiation. I wanted

> to see the Manx Shearwaters, and he took us to the top of a 300-foot headland. ... He fastened a long light rope around a solid boulder and casts it over; then lashed a stouter rope twice around my chest, knotted it with a peculiar knot that climbers know, set himself and my brother and cousin sitting one behind the other with the rope taut between them, and their heels well dug into the turf, and invited me to go ahead.(PS)

He was lowered down the cliff on the end of the rope, and Praeger later commented, 'I think the tyro's chief difficulty is to have full trust in the rope, and have the courage to lie well back, so as to *walk* down or up the cliff'. He emphasised that 'indeed it was necessary to keep on one's feet here, for every ledge was so packed with birds and their eggs and droppings, and the smell was so overpowering, that it was well to be as far from the cliff face as one could'. Despite searching in several burrows he failed to find any shearwaters on his sortie down the cliff, and was hauled safely back to the top. Then, 'Paddy donned the harness and went over instead; when we hauled him up he had a dark bird dangling at his waist'. Following a detailed examination, the shearwater, which Praeger and his companions thought was now surely dead, 'stirred and turned over; and then suddenly it was up into the air and away, and with astonishment we watched its long curved wings bearing it far out to sea'.(PS)

Towards the end of an epic sail around the entire Irish coast in his yacht in the 1990s my father berthed at Church Bay in Rathlin and tied up alongside other visiting boats. During the night he had some trouble with his eyes and, when he mentioned this to the skipper of the neighbouring boat, he was surprised to learn from this man, who turned out by coincidence to be an ophthalmic surgeon, that he was very likely to be getting a detached retina. Speedy action was needed, so the surgeon telephoned the nearest hospital and shortly afterwards a helicopter arrived with medics on board. My father was whisked off to the Royal Victoria Hospital in Belfast, where rapid surgery saved his eyesight.[17]

The cliffs of Rathlin reach their most dramatic at the western end facing the open Atlantic. After leaving the ferry in Church Bay, the only village, I walked about six kilometres to reach the cliffs. Here it is possible to get panoramic views of the stunning seabird colonies, including the ever-popular puffins, that occupy these cliffs. For most of the year, puffins feed far out at sea, returning to land in April. Most start breeding when they are five years old and often live for more than twenty years. Some young, inexperienced birds may change mates after breeding failure, but most will mate with the same partner for many years. During the summer these comical birds share the cliffs near the island's west lighthouse with thousands of other seabirds, from kittiwakes to fulmars, but the huge population of common guillemots (estimated at over 130,000 individuals in 2011[18]) seems to cover every available square metre of ledge, as well as the flat tops of all the sea stacks. This is the largest concentration of this species in Ireland or Britain. From the West Light Seabird Centre, which is run by the RSPB, I got great views of these birds using my binoculars. As well as the visual spectacle, the sounds and smells of the colony were all part of my experience. Constant commuting of birds in and out of the cliffs leaves the impression of a seabird city with a large and busy population.

A new five-year project has recently been announced to boost the seabird population on Rathlin Island, which has dropped by more than 50 per cent in recent years due to predators such as rats and ferrets. Rats are thought to have come to the island via boats, including from shipwrecks, while ferrets were introduced to manage rabbits but escaped and bred. In charge is Claire Barnett, manager for the RSPB, who said the aim is to remove all the ferrets and rats from the island. 'It has worked in the Scilly Isles and it's worked in New Zealand so hopefully it will here too.' She said she hoped increasing the seabird population will also attract more tourists to the island.

The location of the Ulster coast and the influence of the Gulf Stream make Rathlin a meeting place for both northern and southern marine species, although, with climate change, these boundaries may also vary. For example, the sea anemone *Parazoanthus axinellae* reaches its northern limit at Rathlin, while another boreal or arctic species, *Phellia gausapata*, is not found further south than the Antrim coast.[19] The deep waters around the base of Rathlin's cliffs and the fast tidal currents both contribute to the richness and diversity of marine life here. The massive cliffs around Bull Point continue downwards below the waterline, where they are encrusted with colourful animals growing densely over the rock surface. Here there are boulder and cobble habitats that support rare and diverse communities of branching sponges, hydroids (sea firs) and bryozoans (sea mats), all of which provide an important food source for many marine invertebrates, including nudibranchs. Both king scallop and queen scallop are also found living among the boulders, cobbles and adjacent sand on the seabed. A number of species occur here that are rare in Northern Ireland, such as a sea cucumber, a sea sponge and red algae. Recent dive surveys have revealed a stunning thirty species here that are entirely new to science.

However, before Rathlin could be given any environmental protection, the first signs of seabed damage from scallop dredging were found, with photos from divers showing overturned boulders alongside mutilated sponges. Even with the designation of the island under European legislation, divers surveying the seabed after dredging activity in 2009 captured images of trawl scars and displaced invertebrates, including sea pens and a soft coral known as 'dead man's fingers'. Eventually regulations were enacted, making it an offence to use bottom-towed fishing gear around the island within the boundary of the Rathlin Marine Conservation Zone.

Since the ban was imposed in 2016, new species and habitat data have been gathered from within the Rathlin Marine Protected Area through both statutory and recreational recording initiatives. This was compared with information from the 1980s, before scallop dredging began. The introduction of the dredging ban has produced signs of recovery of faunal communities on the seabed around Rathlin, with recent records of 371 species, seventeen of which are included on the current Northern Ireland Priority Species list.

Fair Head

Sailing back from northern waters past the eastern end of Rathlin Island, I had previously been impressed by the dramatic scale of the Fair Head cliffs, clearly visible from Scotland, and which mark the most north-easterly point of the mainland of Ireland. I walked along the cliff path staring down at the tide which rushes out through this northern gate of the Irish Sea. This headland, formed by the typical basalt columns, is capped by a large intrusion of dolerite, another hard, volcanic rock. Weathering and ice over millions of years have hacked away at this sharp corner of Ireland and left beneath the cliff a massive scree slope, or talus, which drops steeply to the sea. Praeger knew this area well, and his climbing skills again came to the fore in this description:

> From Murlough Bay, Fair Head is in full view, with its flat top, its clean-cut gigantic wall of columns, over three hundred feet high, and its wild talus descending another three hundred feet into the ocean. ... The scramble round the base of this great semi-circular rampart, over the talus, which from the summit does not look formidable, is a wonderful experience, but not to be undertaken unless you have plenty of energy and time; for the ruin of the cliff is on a gigantic scale, sloping at a steep angle, and made up of fallen blocks which may be as large as cottages. You climb along, sometimes over, sometimes under, sometimes between these, using both arms and legs to their full capacity, and unless you are a demon of energy, you will be glad when the talus gives out and you see in front the low grey sandstone cliff, and the beginning of the road which leads from the old coal-workings towards Ballycastle.(PS)

His exploits in pursuit of Ireland's natural history show that Praeger was indeed a 'demon of energy' in his physical explorations, his organisational skills and his prolific writings. There were few places in Ireland that he did not explore in his long and active life, and his knowledge of the coastline was exceptional.

Leaving the North Coast

Fair Head marks the end of my rambling account of the Irish coast, but I will never stop exploring and seeking out new and interesting places along the shore and out at sea. The north coast is the shortest stretch of the four coasts, but its interest rivals the other, longer parts. The ancient rocks of north Donegal contrast with the more recent basalt plateau of Antrim and the low-lying sandy coasts of Derry. The only significant offshore islands are Tory, Inishtrahull and Rathlin. The northern shores are often battered by Atlantic storms, with some historic buildings perched on the cliffs above. The Giant's Causeway is one of the few World Heritage sites in Ireland, and a major draw for tourists.

Much of the coastline of Northern Ireland is fully protected either by statutory designations or by the ownership of the National Trust through its long-running campaign Enterprise Neptune. The essential ecology of the north coast is no different from that of the rest of the island, but it appears to be more highly valued by its people, and this is recognised by the government.

Crossing the Bar

In theory, one could go on forever following the continuous coastline of an island like Ireland again and again and, each time, finding new and exciting places. The Irish coast offers infinite opportunities for exploring that complex fringe where the land and sea merge, forming multiple habitat zones and supporting myriads of species. Our island is so small that one is never far from the coast and, like Praeger, I am always drawn, as if by a magnet, to explore harbours, beaches, dunes, cliffs and islands. The islands off the Irish coastline hold a special attraction for me, as they did for Praeger. He frequently wrote about his excitement on visiting these isolated places:

Islands are always fascinating – particularly if

> they are small. Their aloofness makes a curious appeal .... Picture the romance of approaching, after days of unbroken horizon, an unknown island! We cannot hope for this in our prosaic latitudes, but, all the same the most fascinating holiday that our own country offers is, to my mind, a sojourn on one or other of the little islets that lie off the Irish coast.(BS)

The clearly defined nature of the islands also held a fascination for Tim Robinson, who lived in and mapped the Aran Islands. He wrote that 'the island is held by the ocean as a well-formed concept grasped by the mind'.[1] As Praeger recorded on Rathlin Island and Fair Head, he was a fearless climber and hillwalker in his youth, with boundless stamina and drive. In his early life he was described as 'powerful' and 'full of energy'. His friend Anthony Farrington recounted how 'he was near his sixty-seventh birthday when I met him at the top of Coumshingaun Mountain. I had traversed the southern ridge but Praeger had come up the cliff at the back of the corrie.' He was always learning although, when he was 69, he said, 'I am too old to learn new techniques but I want to know about all these things and so I'll do all I can to help'.[2] In the 21st-century world of complicated technology, I can relate to this feeling of detachment from nature and a dependence on computers and mobile phones, but I have a desire to stay focused on the wild places. Praeger was equally undistracted by the 'new techniques' and his fieldwork continued until he was over 80.

I have often wondered what drove Praeger to devote his entire life to exploring the island of Ireland in such minute detail, inspiring many other naturalists to collaborate with him and resulting in such ground-breaking results as the Clare Island Survey. His early family life was obviously a key launchpad in that his parents gave him the freedom to wander at will, exploring the countryside within easy reach of his home, from the Mourne Mountains to the Antrim coast. Here he honed his field skills and developed a love of outdoor pursuits such as hillwalking, climbing, caving and swimming. His grandfather and his uncle were already well-known naturalists in their own right. They inspired the young Praeger, encouraging him to join the Belfast Naturalists' Field Club, where he learned from many experienced naturalists. He was also a product of the Victorian fashion for natural history pursuits such as botany and entomology that were especially associated in Ireland with the Anglo-Irish community.

It is likely that he was influenced in his youth by Darwin's seminal work, *On the Origin of Species,* which was published in 1859, just six years before Praeger was born. This was the foundation for evolutionary biology, and it must have sparked many questions in Praeger's mind about nature in Ireland, which was little studied at that time. Despite being trained as an engineer, he was fired with the curiosity of a scientist to find out more about his natural environment. His interest in archaeology and antiquities may have been reinforced by the growing fascination with Celtic legend and mythology in late 19th century Ireland and the rise of cultural nationalism. In adult life he was so busy with his scientific work that he does not seem to have had time to raise a family. A key factor in all he undertook was the support of his devoted wife Hedwig, who explored many of the remote places with him. Equally, he loved to explore alone and wrote in *The Way that I Went,* 'the study of nature in the open needs contemplation and quiet, and these are becoming increasingly difficult to obtain'.

His penultimate book, *The Natural History of Ireland*, was widely regarded as falling short of the unique nature of his early efforts. Written during World War II, when the author was nearly eighty, but not published until 1950, it suffers from 'too much technical terminology and obscure botanical categories'. It also 'indulges some of the author's pet topics without consideration of their interest for a potential readership'.[3] I recognise from my own experience as a writer how easy it is to become absorbed by the finer details of my own special interests, whether it is sailing or nature study, and forget that these pursuits may of little interest to a wider readership.

Seán Lysaght, who wrote an authoritative biography of Praeger, recalled, 'as an old man he wrote a little book on *Irish Landscape*, in a series commissioned by the Cultural Relations Committee.[4] Máire Mhac an tSaoi, then working for the Department of Foreign Affairs, reviewed the typescript with him and was obliged to write her comments and questions, as Praeger was very deaf and, by now, had not long to live.' Reflecting the policy of the government to consider Ireland as a single entity, she insisted on removing references to 'partition of Ireland' because it 'spoiled the otherwise happy character' of the text. She also asked him to consider changing the references to 'the British Isles' and 'Londonderry' to 'these islands' and 'Derry' respectively. Clearly, Praeger agreed to the censorship, but he was still allowed to include a rant about the untidy appearance of Ireland's towns and villages.[5] The tiredness of old age shows clearly in the rough scrawl of Praeger's handwriting on the manuscript of the *Irish Landscape*, which was eventually published as his last book in 1953.

Anthony Farrington wrote that when Praeger sent his last scientific paper to the press in 1951, he said to a friend, 'I'm written out. I have no more to say,' adding characteristically, for he did not relish idleness, 'can you think of anything that I can do?' Farrington also referred to Praeger's 'somewhat abrupt manner of speech' in old age, saying,

> this habit of 'blustering,' as he himself called it, caused many who did not know him well to regard him with awe and some to call him rude. For this reason, on Field Club excursions the more timid, desiring information about some plant, were wont to approach him through the

> medium of his wife. It must be admitted that he could be very short with those he thought insincere or inept, though once he was convinced of the real interest of the inquirer in the matter concerned, he would go to great trouble to give them the required information. There must be innumerable instances of his kindness. One such is a series of letters written between 1943 and 1945 to a girl of 13 or 14. Here we find Praeger at his kindest, answering questions, advising, or merely chatting. It has been suggested that his bluff manner may have developed early in life to cover a sentimentality which he undoubtedly possessed and of which in his youth he may have been shy. In his later days he was not ashamed to admit this by various actions known only to a few of his most intimate friends.[6]

With advancing years, Praeger found it hard to adapt to modern life. 'The present time is one of rush and clatter, of fuss and noise and glare: I fancy I see repercussions of all this or of the new mentality which has produced it, and which it has produced, in the strange literature and art and music of the day.

My medieval mind will not rise to modern heights'. Like him, I sense a need for peace and quiet as I grow older, finding solace and escape in sailing far from the distractions of the modern world or absorbed in watching a wild animal or bird as it goes about its simple struggle to survive. Praeger disliked the modern pressures affecting some lifestyles that were largely unchanged in Ireland since the previous century:

> Hurry and noise are the keynotes of today, and where these prevail we need never hope to lure the fairies from their hiding places. Perhaps I have arrived at the stage of old-fogeydom, for I recall the more leisurely, deliberate, spacious days of Queen Victoria, the courteous quakerish naturalists who taught me the truths that the lie at the bottom of all science, and I confess that I look back on those times ... with a deep affection. (WW)

After more than fifty years of marriage, Praeger's wife Hedwig died in 1952. Now aged 87 and almost completely deaf, Praeger had become heavily dependent on his wife for basic needs of mobility and communication with others. After sorting out his affairs and transferring his huge library of papers to the Royal Irish Academy, he moved back to the place of his birth in County Down to live with his sister, Rosamund, who by this stage was a well-known artist. It was there that the great Irish naturalist died in May 1953, aged 88 years. He left behind him an enormous legacy of publications, knowledge of the flora, fauna and archaeology of his native island and a much greater appreciation of the natural world. Sixteen years earlier in *The Way that I Went*, which was subtitled *An Irishman in Ireland*, he had written:

> To the patriot, the loveliest country is – or should be – that in which he was born and in which he has lived, for it has given him the very foundation of his being. ... I have wandered about Europe from Lapland to the Aegean Sea but have always returned with fresh appreciation of my own land. I think that is as it should be.(WW)

In my own wanderings around the Irish coast over the last seven decades I have, in many cases, revisited the same places where Praeger 'roamed at random' up to a century earlier. I have noticed great changes on many wild shores, but others remain much the same as when the greatest Irish naturalist of his time recorded their features.

Turning the Tide

Despite living through tumultuous political times from the late Victorian era through the fight for Irish independence and both world wars, Praeger could not have imagined the dramatic changes in Ireland that would follow his death in 1953. At that time, the majority of Irish people lived in the countryside. Few people owned a private car or could afford to take a holiday other than a few days by the seaside in Ireland. About one-third of Irish homes still had no electricity or running water. Agriculture was the main form of employment and the horse was the 'tractor' of the early 20th century. Transport to the many offshore islands was still mainly by rowing or sailing boat and the oceans seemed to hold a limitless supply of fish. Protection of nature was rarely discussed except in the context of hunting, fishing or game management. Praeger hardly ever mentioned threats to nature in his writings, although he was aware of some small changes happening in the early 20th century. On the north coast, where he had once played as a child, he noted 'the sand-dunes, now thronged with golfers, were then lonely and exciting places'.(WW)

Even in the fifty years since I started work in nature conservation, the threats have escalated in scale and significance. When I began working on the sand dunes in Northern Ireland during the 1970s, the main threats to the coast at that time were considered to be oil pollution, invasive species and disturbance from recreational use. While these pressures can still have localised impacts on coastal habitats and wildlife, my ideas have changed. Half a century ago I was not aware that the rise in greenhouse gases, already being detected at that time, was the result of human activity and would have such major implications for the planet. Nor did I realise that fish stocks in Irish waters were being exploited to the point where some of the previously common species would collapse. Recently, I have become aware of the major problem of plastics in the oceans and the long-term impacts of this. All of these problems have been caused by humanity, and it is up to us to fix them.

Dwindling Fish Stocks

I admit to being a seafood fan. A plate of fresh crab, mussels, prawns or mackerel would suit me for dinner anytime. We have such a wonderful choice of locally caught seafood in this country it is a surprise that it is not constantly on the menu of every home, café and restaurant. As a boy, I remember being invited to go out in a small wooden rowing boat off the coast of West Cork to join a couple of local men fishing for mackerel at dusk. As the sky darkened, we reeled in scores of fish, and when we finally returned to the pier we had a full sack of the glistening green-and-black-striped mackerel, so many that we could not give them all away.

But all is not well today in the fishy world. Multiple pressures on the marine and coastal ecosystems have intensified since Praeger's time. I sometimes go into the Natural History Museum in Dublin just to look into the glass cabinets at the giant specimens of turbot, brill, plaice, sole and shark that were once caught in the Irish Sea by the crews of sailing boats in the 19th century. Due to overexploitation, fish of this size and variety are absent today. These were caught by line fishing from small wooden boats and later with nets following the introduction of sailing trawlers to Dublin around 1818. An account of a fishing trip off the Dublin coast on one of these trawlers, *The Perseverance*, was published by Dr William Corrigan:

> The produce of the trawling-net, when turned out on the deck, is certainly a curious and interesting sight: comprising skate, ray, brett, turbot, conger eel, John Dory, gurnard, red and gray, cod, haddock, soles, plaice, herrings, mackerel, flounders, squids (small cuttle fish), and piles of queer things about which naturalists get enthusiastic, such as sea-mice, star-fish, sea urchins, comatulæ (feather stars), brittle stars, seaweeds of beautiful colours and forms, beroæ, &c. All these, of no use to the trawler's crew, they call 'curioes.'[1]

By 1900 there were about nine big steam trawlers based in Dublin, and the last of the sailing trawlers ended its days in 1918, having become outdated.

This represented a quantum leap forward in fishing methods by comparison to what had gone before, and led to a consequent diminution of the fish stocks. Overfishing has since decimated many of the fish and shellfish populations of previous centuries, and this is having knock-on effects throughout the marine ecosystem, from prawns to porpoises. The Irish Sea is split between Ireland and the UK, and has been overfished by both as well as by other EU fishing fleets. Overfishing in the Irish Sea has driven the collapse in some commercial fish stocks. The reform of the EU Common Fisheries Policy has seen a reduction in the fishing mortality in some stocks such as plaice and haddock, and these have shown some signs of recovery, but cod and whiting populations are still in a very poor state due, in part, to high levels of bycatch in the prawn fishery.

One of my favourite seafood restaurants is on the west pier at Howth in north Dublin, where I have often sat watching the big trawlers unload their catches in the harbour. Box after box is swung ashore to be loaded into refrigerated trucks and driven away, often to other European markets. However, despite the European Union's legal deadline to end overfishing by 2020, many of the Total Allowable Catch (TAC) limits still exceed the best available scientific advice, facilitating overfishing. BirdWatch Ireland has calculated that over 50 per cent of the TAC limits, of which Ireland has a share, exceeded scientific advice in 2020. According to the Irish Marine Institute, only 20 per cent of commercial fish stocks in Irish waters analysed in 2019 met the criteria for achieving Good Environmental Status. Over the last twenty years, Ireland is among the five EU countries to receive the highest percentage of fish quotas *above* scientifically advised levels for sustainable limits. The wasteful practice of discarding unwanted low-value bycatch at sea also continues, despite an EU legal requirement that all catches be landed to reduce bycatch through increased selectivity. There is evidence to suggest that the so-called 'Landing Obligation' has not been fully implemented in Irish waters, and illegal discarding at sea continues despite the EU's legal deadline of January 2019. Illegal discarding is contributing to ongoing declines in overfished stocks.[2] Although some seabird species like gannets and fulmars have benefitted from exploiting the bycatch, this is not a natural situation, and needs to be changed.

I spent a week living aboard my yacht, tied up in Kilmore Quay Harbour in Wexford, as I waited for a replacement engine part to arrive. Every morning around 5 a.m. the big trawlers in the harbour would drop their lines and steam out to the Celtic Sea to catch more fish. I saw large sharks and tuna being landed here among the more common haddock and cod. Close to the harbour a large factory processes shellfish for export, including thousands of scallops each day.

Predatory fish are not immune from the pressures of overfishing. These species fill ecological niches that are important in maintaining a balance within the ecosystem. Their removal can result in cascading effects that have a negative chain reaction on marine biodiversity right down the food chain. Irish waters contain seventy-one cartilaginous fish species (sharks and rays), over half of the European list. Of these, fifty-eight were assessed using the latest international categories. Six species were considered to be critically endangered – Portuguese dogfish, common (blue) skate, flapper skate, porbeagle shark, white skate and angel shark. For example, numbers of angel sharks recorded in tagging programmes show a decline of over 90 per cent since the 1980s. A further five species were assessed as endangered, while six more species were rated as vulnerable. While there are no longer any vessels fishing directly for threatened cartilaginous fish in Irish waters, some are taken as bycatch in several fisheries, involving both Irish and non-Irish boats. Similarly, endangered and threatened species that straddle Irish and non-Irish waters are caught by fleets further afield.[3] Ireland is not unique in this as, since 1970, the global abundance of sharks and rays has declined by 71 per cent owing to an eighteen-fold increase in relative fishing pressure.[4]

I have been in many small west coast harbours waiting for boats to take me to the islands. Rosbeg Pier in west Donegal is exposed to the full rigours of the Atlantic, and only a few small inshore boats tie up there. Pot fishing for lobster and crab are the main activities, since net fishing for salmon has been banned. At one time piers like this one would have been packed with small wooden boats as local people joined in the harvest of millions of migratory salmon, herring or mackerel. In Praeger's day, many coastal communities were wholly dependent on healthy fish stocks to sustain them over the year. The fishing was almost entirely in inshore waters close to the coast using traditional rowing boats such as curraghs, hand-made wooden lobster pots, long lines and nets laid from wooden sailing boats and latterly by sailing trawlers imported from England. Vast shoals of herring migrated along the Irish coast and large fleets of migrant fishermen followed them to catch the 'silver darlings'. I can remember taking home plenty of herring from West Cork in the 1960s, but I would be very lucky to catch any from my boat today.

While it is easy to blame overexploitation alone for the decline of fish stocks, fishermen have understood for centuries that populations can fluctuate due to natural changes as well. In the 19th century whole communities lived by harvesting herrings and pilchards, but they understood that the peaks in these species can alternate, and that one might be replaced by the other. Herring predominates when one of its key prey, a particular species of arrow worm, occurs in large numbers in the plankton. When environmental conditions, such as the North Atlantic Oscillation, change so that a different species of arrow worm predominates, then pilchard outnumber herring.[5] So if environmental conditions become less favourable for herring, continued targeting of this species will exacerbate the natural fluctuations, making it less likely that stocks will recover.[6]

Supertrawlers can scrape the seafloor bare of bottom-dwelling species and capture huge quantities of shoaling fish like sprat and mackerel in a few hours. Small boats, which make up 80 per cent of registered fishing vessels in Ireland, are totally dependent on stocks in inshore waters. Alex Crowley, fisherman and secretary of the National Inshore Fishermens' Association, told me, 'if you go back ten, fifteen years we had a very diverse inshore fleet then. It supported a lot of small coastal communities. But there are no fish in the bays any more worth fishing for. The herring stocks we have in Ireland have collapsed. The mackerel is the same. Not only has our sector shrunk, it has also become less resilient and more dependent on fewer stocks. Overfishing has contributed significantly to that.'

I visited Northern Ireland's major fishing port of Kilkeel in County Down to see for myself what a modern trawler base is like. After the accession of Britain and Ireland to the EU, Kilkeel's fleet grew from forty boats to more than 160, with local memories still fresh of being able to walk from boat to boat across the full width of the harbour. The fish quotas were largely given to the big corporate-owned offshore fleets with large trawlers. Bottom trawling is used to repeatedly scrape the seafloor with a weighted net which collects scallops, prawns, cod, monkfish, haddock and various flatfish, as well as a confusing mix of non-target species that are treated as bycatch. Repeated trawling of the same areas of seabed can reduce marine life in soft sediments by up to 80 per cent, and represents a major threat to the ecosystem. Bottom trawling is very widespread in the Irish Sea, and especially in its northern muddy areas, which contain the main stocks of Dublin Bay prawn. With repeated trawling up to twelve times a year the abundance of bottom-dwelling invertebrates decreases rapidly. Trawling for shellfish in shallow coastal bays on the west coast is also destroying some of the unique seagrass beds and maërl formations that have taken centuries to develop.[7]

Dredging also has impacts on shellfish in shallow waters. In just a few years a razor-clam bed discovered at Gormanstown, County Meath, was wrecked by fishing boats using custom-built hydraulic dredges to scour the long-lived clams out of the fine sand. By 1999 about two-thirds of them were gone, mostly sold to Spain.[8] Hydraulic dredging to extract razor clams disturbs sediment to a depth of twenty-five centimetres and considerable volumes of sediment are mobilised during fishing. The Marine Institute reported that the impact of this type of fishing on seafloor integrity is significant, especially destroying sediment structure and associated fauna. Some of this activity occurred within the protected area of Dundalk Bay.

In the first eight winters of the 21st century, Dundalk Bay held the only internationally important population of oystercatchers in the country, with a peak of over 15,000 birds in winter 2006–07.[9] I joined the team of ornithologists censusing these waders here for the Marine Institute. By 2015–16 the numbers of these distinctive black-and-white shorebirds had declined dramatically. Common cockles are their most important prey, but a cockle fishery in this protected area had already removed over 1,000 tonnes of these shellfish in just three years. Oystercatchers changed to feeding on other prey or moved completely out of the bay to feed in fields.[10] Nationally, the total numbers of these waders dropped by over a fifth in the same period, as fewer birds spent the winter in estuaries.[11]

One of my pleasures on a visit to Galway Bay is to sit outside Morans of the Weir at Kilcolgan with a pint and a plate of delicious flat oysters in front of me. Morans is a family business dating back more than 250 years. In the 1800s it flourished when this was a prosperous little port. The Weir was named after an old wall constructed across the nearby tidal Dunkellin River to trap salmon. Local people also dredged for Galway Bay oysters from the nearby Clarinbridge oyster beds. At one time almost every bay and many offshore banks had natural beds of the native oyster, so abundant that they were thought to be inexhaustible. But by the 19th century, this renewable resource was decimated and the few small populations remaining on the west coast were threatened by poaching, by an oyster parasite called *Bonamia ostreae* and by poor water quality. From the 1970s, cultivation of Pacific oysters took over from the native species and long lines of oyster trestles in many bays and estuaries became a common sight. By 2004 production of this introduced species exceeded 11,000 tonnes in 250 separate oyster farms.[12] However, there has been little change in the volume of oysters nationally, with 10,500 tonnes produced in 2019, 85per cent of which was exported.[13] The trestles take up a large space on the sandflats that would be naturally used for feeding by shorebirds. A recent study found that knot, sanderlings, dunlins, godwits and ringed plovers tended to avoid the oyster trestles, as they mainly feed in large flocks of tightly packed individuals.[14]

The survival of the native oyster may be further complicated by the fact that feral populations of the introduced Pacific oyster are now well established in a number of bays in Ireland. At first, it was thought that the introduced species, originally from the tropical waters of the Pacific, would simply be reared to maturity here in the extensive intertidal oyster farms and could not breed in cold Irish waters. However, some are now producing spat as seawater temperatures increase and this is leading to the establishment of new oyster reefs in bays such as the Shannon Estuary, Galway Bay, Lough Swilly, Lough Foyle and Strangford Lough.[15] The spread of the introduced species is much more extensive in some other European waters, and evidence has emerged in the North Sea suggesting that the return of native oysters may actually be facilitated in some places by the spat settling and colonising reefs of invasive oysters at sites where their distributions overlap.[16] The indirect effects of introducing species from other parts of the world to our coasts are often hard to predict.

I love to wade out with a bucket to the middle of the Gweebarra estuary in west Donegal at extreme low tide, where the rocks are encrusted with colonies of large blue-black mussel shells. I pull back the curtains of brown seaweed as small shore crabs scuttle away to safety among the barnacle-covered rocks. It takes some strength to prize the mussels off the rocks, as their strong byssal threads bind them to the other shells in the colony. But most mussels sold today come from mussel farms such as those in Killary Harbour, Bantry Bay or Wexford Harbour. This type of aquaculture is normally dependent on the natural 'fall' of spat or juvenile mussels from the plankton reaching the ropes that hang in the water. In Wexford the mussel industry requires large quantities of small seed mussels to be relaid in the shallow waters of the harbour. However, repeated dredging of the Irish Sea for seed mussels has reduced this annual harvest by two-thirds, so that there are now insufficient stocks of these juvenile shellfish to support the economically valuable mussel cultivation industry in Wexford Harbour. The decimation of fish and shellfish populations has serious impacts right through the marine food webs.

One of the most interesting seabird locations I have visited is the kittiwake colony that breeds in and around Dunmore East in Waterford. At one time, on the cliffs that surround the busy harbour, these small ocean-going gulls would raise their chicks within metres of the human activity of a busy fishing port. Today, however, this population of kittiwakes is struggling to produce enough young birds, as the fish species on which they depend become scarce. This is our most abundant breeding gull, but the numbers of adults attempting to breed in Ireland have declined by 35 per cent since the 1980s.[17] The practice of pair-trawling for spawning inshore sprat has increased in recent years, and the existence of these fisheries operating within the foraging areas of kittiwakes and other seabirds has serious implications for the breeding success of species that depend on sprat. A recent study of breeding kittiwakes at two colonies, Rathlin Island and Lambay, showed that breeding success suffered, most likely due to shortages in food resources near the colonies, with adult birds responding by flying further to find food, resulting in greater losses of their eggs and higher numbers of starving chicks.[18]

Overfishing is changing the entire balance of the marine ecosystem. By removing predatory fish from marine systems and then fishing down the food web, fisheries have been transitioning to smaller plankton-feeding fish, including sandeels, young mackerel, sardines and sprats (collectively known as forage-fish). When fisheries target these smaller fish species, it not only puts pressure on seabirds that depend upon them but affects food abundance for the larger predatory fish species, with the overall effect of hindering their recovery and unbalancing the entire marine ecosystem.[19]

For me, one of the highlights of sailing off the West Cork coast is the strong chance of seeing a large whale or a group of dolphins close to the boat. On one memorable occasion, I was accompanied by a pod of common dolphins in crystal-clear waters off Toe Head, and I was able to film them just metres from the bow of the boat. Several whale-watching businesses have been developed based in the picturesque harbours of Baltimore and Castlehaven. In a welcome development, large whales seem to be returning to Irish waters as global populations recover from centuries of excessive hunting. Over a hundred individual humpback whales have been identified by the Irish Whale and Dolphin Group in recent years. Together with fin whales, these depend heavily on small shoaling fish such as sprats and juvenile herring in inshore waters.[20] Historical records tell of the vast herring shoals that used to fill bays and harbours like a great silver tide. John Molloy described large winter shoals of herring that would gather off the Stags of Broadhaven, County Mayo, during the 1960s and 1970s. However, these were heavily exploited and disappeared during the 1980s.[21] Today, three out of the five herring fisheries, in which Irish fishing boats are active, have collapsed. This includes two of Ireland's most important herring fisheries, in the Celtic Sea and north-west of Ireland. Sprat are now under serious pressure too. Worryingly, a ban on fishing by large trawlers inside the six-mile limit was overturned by the Irish courts recently. This ban was designed to protect inshore fish stocks, such as sprat, which is a keystone species, providing important food for a range of predators, such as whales, dolphins, seabirds and sharks, and also a range of commercial fish species. Sprat is too important a species in the marine ecosystem to neglect and merely grind into fish meal.

Should consumers avoid seafood altogether to help stocks recover? This might have some impact in Ireland, but the bulk of the catch here is exported either by Irish boats or by vessels from other countries, so it would take a boycott right across the EU to have any significant impact. There are European laws to prevent overfishing based on scientifically based total allowable catch and to prevent discarding of bycatch at sea, but these are routinely ignored by member states, including the Irish government. Now the UK is outside the EU and so not subject to European laws. The fishing industry, especially the large trawler owners, have a powerful lobby that puts pressure on politicians whenever the Common Fisheries Policy is up for renewal. Even the High Court judgment preventing large trawlers operating within the six-mile limit has been overturned recently, leaving stocks decimated and small inshore fishermen without a proper livelihood. A major campaign by combined consumers and environmental groups across the EU and UK, on a par with recent climate action demonstrations, needs to be mounted to pressure the governments to properly enforce European regulations.

Water Quality Concerns

Where I swim on the sandy beach of Brittas Bay in Wicklow the water is usually crystal clear, and I have no concerns about water quality. In Praeger's time too water pollution was rare and localised but, where it occurred, it was usually the result of untreated sewage entering rivers and estuaries. By the late 19th century Dublin Bay was so badly polluted that consumption of shellfish collected here often caused typhoid fever or death.[22] Even today, pollution incidents frequently lead to the closure of swimming beaches around Dublin. The cause is often the malfunction or overloading of wastewater treatment works such as the giant sewage plant at Ringsend in Dublin Port.[23] Raw sewage from thirty-five towns and villages throughout Ireland continues to be discharged into rivers, estuaries and the sea, all of which ends up on the coast. A report by the Environmental Protection Agency indicates slow progress in developing modern wastewater treatment works to deal with this problem.[24]

Nor is Ireland immune from the problem of plastics throughout the world's oceans. Walking on any beach, even in remote rocky coves of the west coast, I frequently come on concentrations of plastic containers, polystyrene packaging, polypropylene ropes and discarded fishing gear wrapped in seaweed on the strandline. My workshop at home is filled with coils of rope that I have collected on remote west coast beaches far from any harbour. The overwhelming bulk of litter found in beach surveys throughout Ireland is plastic, and this does not decompose but simply accumulates over years.[25] In 2019, a team of researchers from University College Cork, investigating deep-water corals 320 kilometres from land, found plastic rubbish in a two-kilometre-deep submarine canyon in the Porcupine Basin. While operating a remotely operated vehicle equipped with cameras, the team encountered a range of rubbish including black plastic bags, fishing gear and water bottles.

Whales, dolphins and porpoises (cetaceans) and most seabirds feed in the ocean, so they too are exposed to plastics. Researchers in Galway Mayo Institute of Technology (GMIT) and University College Cork (UCC) examined a large number of stranded cetaceans, of which 251 had signs of possible entanglement or interactions with fisheries. In post-mortem examinations of 528 stranded and bycaught cetaceans, forty-five animals had marine debris in their digestive tracts, with a large proportion being fisheries-related items. The release of microplastics into the sea from washing artificial fabrics is now a major area of concern, as the microscopic particles clog up marine food chains and also enter the human food chain as a contaminant in fish and shellfish. All twenty-one cetaceans investigated for these pollutants in the GMIT/UCC study contained microplastics, with the majority being composed of fibres.[26] Heidi Acampora of GMIT investigated the amount of plastic pollution in dead seabirds found stranded on beaches in Ireland. Among fourteen fulmars dissected, all but one had plastics in their stomachs, with the average bird containing sixty-five plastic particles, weighing 1.1 grams. Plastics were also found in several other seabird species, such as guillemots and gannets.[27] There is a general lack of awareness among the public about the adverse effects of micro-fibres originating from washing of artificial fabrics such as fleeces. A campaign of action is urgently required.

If consumers understood the long-term damage being done by plastics in the oceans they might accept a levy on the sale of items such as single-use disposable containers, just as Irish consumers did with the 'plastic bag levy'. But a high proportion of the plastic waste originates with discards from the fishing industry, and this is very hard to control. Equally, plastic washing up in Irish waters may have originated from other countries and even from places outside Europe. Unless bans on the sources of the problem and practices such as dumping at sea are properly implemented this is likely to continue.

Climate Crisis

Like most people, I tend to remember the hot summer days of childhood and carefree holidays spent on the beach or messing about in boats, while choosing to forget the long cold or wet years when we felt cheated of our holiday fun. We are fortunate in Ireland in that we have few extreme temperatures, with our western shores washed by the Gulf Stream (or North Atlantic Drift) to keep us warm in winter. I know people who swim in the sea almost every day of the year and love it. Our weather changes from day to day and from week to week but the climate is subject to much longer patterns and trends. Climate change is already having significant long-term impacts on the oceans, on coastal habitats and wildlife. Since the middle of the 20th century, the oceans have absorbed roughly 93 per cent of the excess heat caused by greenhouse gases from activities such as burning coal for electricity. That has shielded the land from some of the worst effects of rising emissions. However, the rising temperature of seawater is leading to expansion of the oceans and the retreat of cold-water species further north. Globally, this temperature rise has contributed to mass coral reef bleaching, the loss of critical ecosystems, and threatened livelihoods like fishing as the target species have moved in search of cooler waters. Best estimates of ocean warming in the top 100 metres are about 0.6 to 2.0°C and about 0.3 to 0.6°C, at a depth of about one kilometre, by the end of the 21st century. However, due to the long timescales of this heat transfer from the surface to depth, ocean warming will continue for centuries, even if greenhouse gas emissions are decreased or concentrations kept constant.[28]

I have stood on the coastal promenade at Clontarf on a spring high tide and watched the waves from Dublin Bay lapping around park benches and spilling onto the Howth Road around parked cars. Sea level rise is a real issue for the residents of these low-lying parts of some of our major cities. Expansion of the oceans combined with the melting of glaciers worldwide is causing sea level rise and, along with increased frequency of storms, we are facing more severe coastal flooding, increased erosion of soft beaches and dunes and coastal squeeze. This means that the low water mark is moving closer to land and the intertidal zone is being reduced in area, thus depleting the available habitat for a whole range of animals from cockles to curlews. Recent reports suggest that sea levels worldwide are rising faster than predicted, and experts on coastal flooding say that 'our beachfront buildings are threatened as never before. It now comes down to buildings or beaches: we must make our choice.'[29]

The interplay of marine and terrestrial forces is graphically illustrated by a mobile dune system that I know at Loughros More Bay in west Donegal. When I first went here in the 1980s the waves reached almost to the doors of a cluster of mobile homes at the top of the beach. Today, the caravan dwellers cannot even see the sea, and they have to walk through hundreds of metres of new sand dunes to reach the beautiful beach of Trá More. The coast here is quite mobile, as the entrance to the bay moves around with sand eroded on one side and deposited on the other to be reversed again after a few decades. What was once the seabed is now a well-vegetated dune system and vice versa. The changing climate can impact currents in the sea, and this has knock-on effects on coastal ecosystems such as beaches and dunes.

The coastal zone comprises a combination of ecosystems – marine, intertidal and terrestrial – that are all interconnected and dependent on one another. For this reason, the management of one component can have indirect impacts on others. For example, the installation of hard coastal protection such as rock armoury in front of an eroding sand dune system has the inevitable result of degrading the beach sands as the sea transfers its energy to the intertidal area, and this may even result in undermining of the rock armoury. Sand dunes depend on the supply of blown sand from the beach to maintain the different dune-building grasses which naturally repair eroded areas. When the beach-dune processes are interrupted both dunes and beach suffer.[30] Current estimates are that the rate of erosion for the Irish coast is between 0.2m and 1.6m per annum, with sand dunes and soft cliffs being the worst affected.[31] Currently, no less than 350 kilometres of the Irish coastline are protected by artificial sea walls.[32] Sea walls can also cause more long-term issues, such as erosion in other coastal areas by depriving them of sediments that would have previously accumulated there. Softer, more environmentally friendly solutions with fewer negative impacts include 'beach replenishment', where sand and sediments are transported from offshore and added to beaches following erosion events.

Turning the tide of climate change is a massive challenge for humanity, and it will take a fundamental transformation in our approach to make any improvement in our lifetimes.[33] I firmly believe that, until there are disasters affecting the whole population, such as complete collapse of fish and shellfish stocks due to temperature rise in seawater, flooding of coastal cities and the disappearance of whole beach-dune systems, national governments will continue to prevaricate and the problems will continue. An encouraging trend is the increased awareness of the young generations and their willingness to engage in political action to make big changes happen. In a few decades, they will be the decision makers, but time is running out, and we have to hope that it will not then be too late for action.

Rewilding the Coast

Treating the coast as a dynamic system with space to change naturally is a more enlightened approach than trying to stop the rise of sea level. This type of management has been practised for many years in other countries, and I first became aware that it was needed while working as a warden at Murlough Nature Reserve in Northern Ireland. At sites sites like this, recreational use is carefully controlled to ensure that it does not lead to damage. This type of management which maintains a 'dynamic equilibrium' should be applied on a much wider scale in coastal areas that are not strictly nature reserves but which are used for agriculture, golf links, holiday homes and general recreation. The key to this approach is the ability to relocate different land uses where overexploitation occurs and in response to dynamic changes in the coastal system. A policy of managed retreat will be an important response to the certain increase in coastal flooding that we face.[34]

There is also great potential for the inshore waters around this country to be returned to something of their original richness and bounty. Pádraic Fogarty points to the example set by Norway, which was never part of the

European Union, but still exploited marine resources as much as any other coastal country in Europe. The Norwegians introduced a policy of sustainable fish catches which have transformed their fishing industry. The trawlers in that country were prepared to accept scientific quotas for their catches, to abide by closures of fisheries when needed to allow fish stocks to recover and to back this up with strong regulations. As a result, stocks of commercially harvested fish have multiplied naturally and Norwegian fisheries are profitable again.[35] A lower catch with a higher value can also lead to a more sustainable income for the fishing families.

Norway was also once the leading country in Europe to exploit the whales of the Atlantic Ocean, from the Arctic to the tropics. They even set up whaling stations on the west coast of Ireland over a century ago, but these did not last long. Norwegian boats still engage in whaling, catching about 1,000 minke whales per year in addition to which cetaceans throughout the oceans now face a wide variety of other threats, such as fisheries bycatch and entanglement, overfishing, pollution (noise, chemical and marine debris) and habitat destruction. In 1991, the Irish government declared the coastal waters of Ireland to be a 'sanctuary' where all whales, dolphins and porpoises would be fully protected. However, few practical measures were taken to follow up this laudable aim. There have been further designations of Special Areas of Conservation for cetacean species under European law but meanwhile, the marine ecosystems on which they depend have not been maintained, as many fish populations have been decimated and other marine ecosystems, even within the SACs, destroyed or degraded. The Irish Whale and Dolphin Group says that 2017 was the worst year on record for stranding of dead cetaceans in Ireland.

The need for Marine Protected Areas (MPAs) is urgent and real. Although definitions vary, MPAs can be thought of as marine areas that are managed over the long term, with a primary objective of conserving habitats and/or species and other natural features. The need to expand the network of MPAs in Irish waters is recognised by the Irish government in order to address the current situation, which is not satisfactory. An expert advisory group has concluded that 'Ireland's existing network of protected areas cannot be considered coherent, representative, connected or resilient or to be meeting Ireland's international commitments and legal obligations. There is no definition of MPA in Irish law and this is a gap which needs to be addressed.'[36] Where fisheries, aquaculture, dredging or dumping at sea are considered to be detrimental to the natural balance in a MPA, then these activities should be suspended or should be banned altogether until the ecosystem recovers. A simple measure that would aid fishing vessels in recognising and respecting protected areas would be to show the boundaries of MPAs on the marine charts that are widely used both in paper and digital formats. Conservation objectives and management plans for MPAs need to take account of the historical richness of our seas instead of setting the benchmarks at maintaining a status quo. The plight of the native oyster, which was driven to extinction in most of our inshore waters some centuries ago, is a perfect example. Restoration of this key mollusc to its former habitats would have enormous benefits for the marine ecosystem.[37] If all Irish coastal waters were managed sustainably, as they ought to be, this would, over the long-term, benefit both fisheries and marine conservation and eliminate the need for MPAs.

MPAs are equally important in Northern Ireland, where they constitute an umbrella term for many other protected areas. There are already a range of designated

European sites, Special Protection Areas (for birds) and Special Areas of Conservation (for other species and habitats) which, despite being no longer part of the EU system, currently retain their legal status. In addition, there are a number of Marine Conservation Zones (MCZs), which often overlap with the SACs and SPAs but have been designated for more locally significant species and habitats, and have greater flexibility in how they are managed. Also, significant tracts of intertidal zone and adjacent maritime lands have been designated as Areas of Special Scientific Interest (ASSIs) or National Nature Reserves (NNRs), adding flavour to the 'alphabet soup'. Thus, a considerable proportion of the Northern Ireland coast has some measure of protection – on paper, at least. The key question is how the designated features are faring, a much more difficult issue; data on their condition is sporadic and difficult to interpret. These sites are largely vulnerable to the same pressures as those in the Republic, and consultations about management, particularly of fisheries, are ongoing. Not surprisingly, the fishing community has difficulties with many of the proposals in the consultation, largely because it acknowledges that use of mobile gear (trawls and dredges) is incompatible with conservation of benthic habitats. However, if 'favourable conservation status' can be achieved overall, this would go some way to achieving a coherent network of representative species and habitats. Incidentally, MCZs can also be used to protect archaeological or historical sites and geological features. An opportunity is also presented here by the ability of some ecosystems – eelgrass beds, oyster beds, for example – to fix atmospheric carbon dioxide or 'blue carbon'. Restoration of these habitats could present government with a 'win-win' situation: protecting important habitats and species, supporting some commercial fisheries and helping to achieve net zero greenhouse gas emissions.

A good example of a working MPA is provided by the Lyme Bay Fisheries and Conservation Reserve, a protected area in southern England in which multiple uses such as fishing are allowed, as long as none are damaging to the seabed or to nature conservation. The project has forged links between fishermen, conservationists, regulators and scientists in order to maintain a healthy, productive and sustainable Marine Reserve within the bay that will benefit fishermen and conservationists alike. The Reserve has achieved three objectives: to protect the biodiversity of Lyme Bay, to implement best practice in managing fish and shellfish stocks and to create long-term benefits for coastal communities around the bay. One of the aims here is to help fishermen achieve best quality and top pricing for their catch. To do this they have created the 'Reserve Seafood' brand, which markets the low-impact, sustainable, premium-quality, provenance-assured seafood of Lyme Bay. There is clearly a premium market for sustainable seafood, so each fisherman is signed up to the Lyme Bay Fisheries and Conservation Reserve and is accredited by the Seafish Responsible Fishing Scheme. This scheme assures catch quality and best fishing standards. The voluntary Codes of Conduct that each fisherman adheres to and the science which underpins the results of fishing efforts in the bay inform the sustainability of the product. Each vessel is also fitted with an inshore Vessel Monitoring System which guarantees the low-impact traceability of each catch. All of this helps towards the long-term sustainable future of Lyme Bay for both fish and fishermen.

We are facing into a period of unprecedented climate crisis, and this is at last being partly addressed by proposed establishment of offshore wind farms on shallow submarine banks along the east coasts of Ireland to generate renewable energy. However, this useful measure needs to be undertaken carefully, avoiding traditional fishing grounds, important seabird foraging areas and ensuring that the massive structures do not disrupt the supply of marine sediments which are vital to replenish our beaches and sand dunes.

In the fast tidal currents of the Narrows at the mouth of Strangford Lough I have sailed past a large black-and-red structure which was the world's first experimental, commercial-scale tidal turbine, installed in 2008 but now dismantled. Below the water there were two large propellors, like those of an early aeroplane, which turned one way on the flood tide and reversed on the ebb. By 2012 the tidal generator had produced five gigawatts per hour of renewable power since its commissioning, which is equivalent to the annual power consumption of 1,500 households. From the start, a monitoring programme was set up to assess the environmental impact of the project, especially on marine mammals which inhabit and breed in the lough. The project received full environmental clearance in January 2012 after the monitoring report showed that there were no major impacts on the marine life in

Strangford Lough. Bob Brown, former National Trust manager of the Strangford Lough Wildlife Scheme, believes that this experiment demonstrated the enormous potential of subtidal currents for renewable power generation. He says, 'with the Irish Sea tidal gates at North Channel and St George's Channel, the power from the tide would be strong and, of course, extremely predictable (so long as the pull of the moon remains!). I believe that, with the right technology, submarine installations could be accommodated in such areas, at sufficient scale to be effective, without compromising marine conservation objectives.'

Unfortunately, decision makers in Ireland constantly ignore coastal conservation, focusing instead on the management of land. Despite being a maritime island, the two governments in Ireland have not traditionally given the sea and the coast the attention they deserve. For decades we have discussed and promoted the concept of Integrated Coastal Zone Management but, for some reason, this has never been adopted in any meaningful way by the authorities in Ireland. When it is successful, it treats the coastal zone and inshore waters as a unit to be managed jointly by all the stakeholders. The problem seems to be that there are far too many sectoral interests here, all with different agendas, and they are unwilling to collaborate.

This problem has been partly overcome in other places, such as California, by general collaboration between all the major users of the coast. On a trip to the Golden State I went to visit the headquarters of Point Blue Conservation Science to learn about the San Francisco Estuary Partnership. This is a collaborative regional program of resource agencies, non-profit organisations, citizens and scientists working together to protect, restore and enhance water quality and fish and wildlife habitat in and around the San Francisco Bay Delta Estuary. Working cooperatively, they share information and resources that result in studies, projects and programmes to improve the estuary and communicate its value and needs to the public. I was taken to visit some abandoned salt lagoons that have been reflooded with seawater and act by absorbing wave energy while providing a rich wildlife habitat. Flocks of marbled godwits called as they flew over my head to their high tide roost in the lagoons. At Point Blue, they have developed the concept of climate-smart restoration as a process of enhancing ecological function of degraded or destroyed areas in a manner that prepares them for the consequences of climate change.

The establishment of Dublin Bay Biosphere Partnership in 2015 offered an example of how this might be achieved in Ireland. But this far-sighted initiative now needs expansion to include many other stakeholders, to broaden its funding base and attract a greater buy-in by the state and semi-state utilities that have a stake in the bay. A separate network of non-statutory organisations (residents' groups, water sports clubs, schools and maritime businesses) could be established as voluntary Biosphere Supporters to help implement its actions. It needs realistic financing and professional staff if it is to achieve its objectives of becoming a place that is 'actively managed to promote a balanced relationship between people and nature'.[38] Dublin Bay also provides a template for how to manage other heavily exploited coastal areas in an integrated way. If we can do this successfully we will enjoy enormous benefits of sustainable harvesting of marine resources and the maintenance of key tourist attractions such as the Wild Atlantic Way. From the ageless pleasure of children playing on a sandy beach, numerous watersport activities, the health benefits of a seafood-based diet, to the generation of renewable energy from the marine environment, the entire population will have a resource that sustains us into the future.[39]

A Noble Goal

In his later years Praeger was a strong advocate of nature conservation, and this was recognised when he was elected as the first President of An Taisce, the National Trust for Ireland. 'I take it,' he said in an address broadcast by Radio Éireann in 1948,

> that we are at the beginning of a very long and also delicate piece of work, calling for patience, tact, judgement and industry, as well as enthusiasm; but our goal is a noble one, and once it is fully appreciated there is very little reason that anyone's hand should be turned against us. Of necessity we begin in a very modest way, but by degrees the movement will gain adherence and influence and become an important factor in our national life.

He was then close to the end of a lifetime's work

which added enormously to our knowledge of nature and of the early human inhabitants here, as he explored and studied every corner of this island in great detail.

In reality, the conservation movement in Ireland has been fighting this 'long and delicate piece of work' as a rearguard battle for at least half a century, and during this time there have been substantial changes in coastal resources that add up to a serious loss for our country. I have worked in nature conservation for most of my adult life, with a special interest in the coast. My first job was looking after a coastal nature reserve, and I have since sailed around the coast for decades. I can see huge changes that have affected the coast and the sea in the century since Praeger studied it. The climate crisis is already bringing new pressures and threats, and the science is quite clear now about the importance of the sea in regulating our climate and of its bounty for our daily lives.

A recent, independent, global review on the Economics of Biodiversity proposes a fundamental change in how we think about and approach economics if we are to reverse biodiversity loss and protect and enhance our prosperity. 'Humanity must ensure its demands on nature do not exceed its sustainable supply and must increase the global supply of natural assets relative to their current level. For example, expanding and improving management of protected areas; increasing investment in nature-based solutions; and deploying policies that discourage damaging forms of consumption and production.'[40] It is time for a turning of the tide.

I will never manage to visit every place that Praeger went on the fringes of Ireland. Nevertheless, as long as I live, I hope to go on exploring it by sailing, swimming and walking on the coast, which I regard as one of our greatest natural assets. I love the smell of the salt air and the constant movement of the tide, wind and waves. The coastline and the sea are among the least modified parts of this island of Ireland. Although there are many pressures and threats, the sea has a great capacity to heal itself given space and time to recover. The return of the great whales from near extinction, the recovery of a shoreline after a pollution incident or the rewilding of a dune system following disturbance are all signs that nature is resilient. While most of it is owned by everyone, I prefer to think of the coast as owned by no one but rather shared between us and the rest of nature. We have a responsibility to future generations to leave the coast in a better condition than we found it. I look forward to a time when, as Praeger said on radio in 1948, nature conservation will 'become an important factor in our national life'.

I took a walk around a familiar rocky headland in west Donegal that I have been visiting for at least forty years. In over an hour of walking I saw nobody else, just a flock of oystercatchers roosting in a small cove surrounded by cliffs and a single seal watching me from the sea. I passed by an old stone lime kiln in which local people used to burn dried seaweed from the shore over the centuries and the tell-tale ridges in the coastal fields where potatoes were grown at the time of the Great Famine. I thought of how countless past generations have eked out a living from the land and sea on these remote coastlines in a way that was largely sustainable and respectful of nature. Hopefully we can learn from their example and treat the coast with the respect that it deserves.

References

The text includes many quotations from Praeger's popular books *Beyond Soundings, The Way that I Went* and *A Populous Solitude*. After the first mention, these books are referenced by (BS), (WW) and (PS) respectively.

Preface & Introduction

1 Praeger, R.L. (1937). *The Way that I Went.* Dublin. Hodges Figgis.

2 Praeger, R.L. (1930). *Beyond Soundings*. Dublin & Cork. The Talbot Press.

3 Praeger, R.L. (1941). *A Populous Solitude.* London. Methuen; Dublin. Hodges Figgis.

4 Collins, T. (1985). *Floriat Hibernia: A bio-bibliography of Robert Lloyd Praeger 1865–1953*. Dublin. Royal Dublin Society.

5 Lysaght, S. (1998). *Robert Lloyd Praeger: The life of a naturalist.* Dublin. Four Courts Press.

6 Neilson, B. & Costello, M.J. (1999). The relative lengths of intertidal substrata around the coastline of Ireland as determined by digital methods in a Geographic Information System. *Estuarine and Coastal Shelf Sciences* 49, pp. 501–508.

7 Fraser Darling, F. (1943). *Island Farm*. London. Bell and Sons.

8 Nairn, R. (2005). *Ireland's Coastline: Exploring its nature and heritage*. Cork. Collins Press.
9 Praeger, R.L. (1949). *Some Irish Naturalists: a biographical note-book*. Dundalk. W. Tempest, Dundalgan Press.

East Coast

1 Blaney, R. (1999). The Praeger family of Holywood. *Ulster Genealogical Review* 15: pp. 91–100.
2 Farrington, A. (1954). Robert Lloyd Praeger 1865–1953. *Irish Naturalists' Journal* 11 (6): pp. 141–171.
3 Long, B. (1993). *Bright Light, White Water: The story of Irish lighthouses and their people*. Dublin. New Island Books.
4 Heery, S. (2018). Charles Joseph Patten: professor and Irish ornithologist. *Irish Birds* 11: pp. 101–105.
5 Patten, C.J. (1914). Missel thrushes, Fieldfares and Redwings at the Maidens Light-house, Co. Antrim. *Irish Naturalist* 23, p. 123.
6 Praeger, R.L. (1885). Rambles round the Antrim coast. 1. Black Head and the Gobbins. *Northern Whig* May 30th, 1885.
7 Greenwood, J.D. (2010). Black Guillemots at Bangor, Co Down: a 25-year study. *British Wildlife* 21, pp. 153–158.
8 Greenwood, J.D. (2015). Breeding sites and breeding success in 2014 of Black Guillemots at Bangor Marina, Co. Down. *Northern Ireland Seabird Report* 2014.
9 Long, B. (1993). *op. cit.*

10 McKee, N. (1976). Copeland. In: *Bird Observatories in Britain and Ireland* (edited by R. Durman). Berkhamstead. T. & A.D. Poyser.

11 Brown, R. (1990). *Strangford Lough: The wildlife of an Irish Sea Lough*. Belfast. The Institute of Irish Studies. Queen's University Belfast.

12 Bell, S. (1951). *December Bride*. London. Dobson.

13 Culloch, R., Horne, N. & Kregting, L. (2017). *A review of Northern Ireland seal count data 1992–2017: Investigating population trends and recommendations for future monitoring*. Report to the Northern Ireland Environment Agency. Queen's University of Belfast.

14 McErlean, T., McConkey, R. & Forsythe, W. (2002). *Strangford Lough: An archaeological survey of the maritime cultural landscape*. Belfast. Blackstaff Press/ Environment & Heritage Service.

15 Wilson, I. (1979) *Shipwrecks of the Ulster Coast*. Coleraine. Impact Amergin.

16 Brown, B. (2011). A magic shell: the seas and shores. In: *The Natural History of Ulster* (eds. Faulkner, J. and Thompson, R.). Belfast. National Museums Northern Ireland.

17 Whatmough, J. (2011). The edge of the land: coast and islands. In: *The Natural History of Ulster* (eds. Faulkner, J. and Thompson, R.). Belfast. National Museums Northern Ireland.

18 Evans, E.E. (1978). *Mourne Country: Landscape and life in South Down*. Dundalk. Dundalgan Press.

19 Cabot, D. & Nisbet, I. (2013). *Terns.* London. HarperCollins.

20 Newton, S. (2021). East coast terns. *Wings* magazine no. 102. Kilcoole. BirdWatch Ireland.

21 Praeger, R.L. (1907). Contributions to the Natural History of Lambay. *Irish Naturalist* 16: pp. 1–112.

22 Hart, H.C. (1883). Notes on the Flora of Lambay Island, County of Dublin. *Proceedings of the Royal Irish Academy Series II*, vol. 3: pp. 670–693.

23 Jebb, M.H.P. (no date). The vascular flora of Lambay. Unpublished report. National Botanic Gardens, Glasnevin, Dublin.

24 Merne, O.J. & Madden, B. (1999). Breeding seabirds of Lambay, County Dublin. *Irish Birds* 6: pp. 345–359.

25 Cummins, S., Lauder, C., Lauder, A. & Tierney, T.D. (2019). *The Status of Ireland's Breeding Seabirds: Birds Directive Article 12 Reporting 2013–2018.* Irish Wildlife Manuals, No. 114. Dublin. National Parks and Wildlife Service, Department of Culture, Heritage and the Gaeltacht.

26 Cummins *et al.* (2019). *ibid.*

27 Cummins *et al.* (2019). *ibid.*

28 Cooney, G. (2000). *Landscapes of Neolithic Ireland.* London. Routledge.

29 Hutchison, S. (2013). *Beware the coast of Ireland.* Dublin. Wordwell.

30 West, A.B., Partridge, J.K. & Lovitt, A. (1979). The cockle *Cerastoderma edule* (L.) on the South Bull,

Dublin Bay: Population parameters and fishery potential. *Irish Fisheries Investigations* Series B, no. 20. Dublin. Department of Fisheries and Forestry.

31 Nairn, R., Jeffrey, D. & Goodbody, R. (2017). *Dublin Bay: nature and history*. Cork. Collins Press.

32 Merne, O.J. (2004). Common *Sterna hirundo* and Arctic Terns *S. paradisaea* breeding in Dublin Port, County Dublin, 1995–2003. *Irish Birds* 7, pp. 369–374.

33 Bolton, J., Carey, T., Goodbody, R. & Clabby, G. (2012). *The Martello Towers of Dublin*. Dún Laoghaire Rathdown County Council and Fingal County Council.

34 O'Sullivan, A. & Breen, C. (2007). *Maritime Ireland: An archaeology of coastal communities*. Stroud. Tempus Publishing.

35 Warren, G. & Westley, K. (2020). 'They made no effort to explore the interior of the country'. Coastal landscapes, hunter-gatherers and the islands of Ireland. In: *Coastal Landscapes of the Mesolithic*. (ed. A. Schulke). Oxford and New York. Routledge. pp. 73–98.

36 McQuade, M. & O'Donnell, L (2009). The excavation of late Mesolithic fish trap remains from the Liffey Estuary, Dublin, Ireland. In: S. McCartan, R. Sheulting, G. Warren & P. Woodman. (eds.), *Mesolithic Horizons*. Oxbow. pp. 889–894.

37 Praeger, R.L. (1896). A submerged pine forest. *Irish Naturalist* 20, 155–160.

38 Newton, S. (2021). *op. cit.*

39 Nairn, R. & Crowley, M. (1998). *Wild Wicklow: Nature in the Garden of Ireland*. Dublin. Country House.

40 Drew, F. (2010). *Racing Round Ireland: a miscellany.* Dublin. Original Writing.

41 Long, B. (1993). *op. cit.*

42 Rees, J. (2004). *Arklow: The story of a town.* Arklow. Dee-Jay Publications.

43 Wilson, J.G. (1980). Heavy metals in the estuarine macrofauna of the east coast of Ireland. *Journal of Life Science – Royal Dublin Society* 1: pp. 183–189.

44 Wilkins, N.P. (2001). *Squires, Spalpeens and Spats: Oysters and oystering in Galway Bay.* Galway. Privately published.

45 Wilkins, N.P. (2004). *Alive Alive O: The shellfish and shellfisheries of Ireland.* Kinvara. Tír Eolas.

46 Bord Iascaigh Mhara (2020). Seed Mussel Survey Report for the South Wicklow Head – 12/08/2020 to 10/09/2020. Dún Laoghaire. Bord Iascaigh Mhara.

47 Merne, O.J. (1974). *The Birds of Wexford, Ireland.* Wexford. South-East Tourism and Bord Failte.

48 Long, B. (1993). *op. cit.*

49 Hutchison, S. (2013) *op. cit.*

50 Reynolds, M. (2003). *Tragedy at Tuskar Rock.* Dublin. Gill & Macmillan.

South Coast

1 Lysaght, S. (1998). *op. cit.*

2 Healy, B. (2003). Coastal Lagoons. In: *Wetlands of Ireland.* R. Otte (ed). Dublin. University College. Dublin Press. pp. 44–78.

3 Cummins, S. *et al.* (2019). *op. cit.*

4 Roche, R. & Merne, O. (1977). *Saltees: Islands of birds and legends*. Dublin. O'Brien Press.

5 Perry, K.W. & Warburton, S.W. (1976). *The Birds and Flowers of the Saltee Islands*. Belfast. Privately published.

6 Cummins, S. *et al.* (2019). *op. cit.*

7 Morris, C.D. & Duck, C.D. (2019) Aerial thermal-imaging survey of seals in Ireland, 2017 to 2018. Irish Wildlife Manuals, No. 111. Dublin. National Parks and Wildlife Service, Department of Culture, Heritage and the Gaeltacht.

8 Long, B. (1993). *op. cit.*

9 Taylor, R.M. (2004). *The Lighthouses of Ireland: a personal history.* Cork. Collins Press.

10 McGrath, D. (2011). *A Guide to the Waterford Coast.* Waterford. Privately published.

11 Cummins, S. *et al.* (2019). *op. cit.*

12 Cummins, S., Lewis, L.J. & Egan, S. (2016). *Life on the Edge – Seabird and Fisheries in Irish Waters*. Kilcoole. BirdWatch Ireland.

13 Lewis, L., Butler, A. & Guest, B. (2017). Intertidal habitat creation – Kilmacleague Compensatory Wetlands, Tramore, Co. Waterford. *Irish Birds* 10: p. 622.

14 Trewby, M., Gray, N., Cummins, S., Thomas, G. & Newton, S. (2006). The status and ecology of the Chough *Pyrrhocorax pyrrhocorax* in the Republic of

Ireland, 2002–2005. BirdWatch Ireland Final Report to National Parks and Wildlife Service, Kilcoole, Co. Wicklow.

15 Robertson, J. (2021). Arthur's Crow: the spirit of the wild? *British Wildlife* 32: pp. 269–275.

16 Gittings, T. & O'Donoghue, P. (2012). *The effects of intertidal oyster culture on the spatial distribution of waterbirds.* Marine Institute Bird Studies. Cork. Atkins.

17 Long, B. (1993). *op. cit.*

18 Lincoln, S. (no date). The Wreck of the Nellie Fleming. *The Ardmore Journal.* Waterford County Museum.

19 Long, B. (1993). *op. cit.*

20 Smiddy, P. & O'Halloran, J. (2006). The waterfowl of Ballycotton, County Cork: population change over 35 years, 1970/71 to 2004/05. *Irish Birds* 8: pp. 65–78.

21 O'Mahony, B. & Smiddy, P. (2017). Breeding of the Common Tern *Sterna hirundo* in Cork Harbour 1983–2017. *Irish Birds* 10: pp. 535–540.

22 Thuillier. J. (2014). *Kinsale Harbour: A history.* Cork. Collins Press.

23 Shanahan, M. (2011). *Martin's Fishy Fishy Cookbook.* Cork. Estragon Press.

24 Somerville-Large, P. (1972). *The Coast of West Cork.* London. Victor Gollancz.

25 Van Gelderen, G. ed. (1979). *The Irish Wildlife Book.* Dublin. Irish Wildlife Federation.

26 Norton, T. (2002) *Reflections on a Summer Sea.* London. Arrow Books.

27 Myers, A.A., Little, C., Costello, M.J. & Partridge, J.C. (1991). *The Ecology of Lough Hyne*. Dublin. Royal Irish Academy.

28 McAllen, R., Trowbridge, C., Bell, J. Nunn, J. & Little, C. (2021). Lough Hyne: a marine reserve in crisis. In: *The Coastal Atlas of Ireland* (ed. Devoy, R. *et al.*) Cork. Cork University Press.

29 McCarthy, B. (2012). *Pirates of Baltimore: from the Middle Ages to the Seventeenth Century*. Baltimore Castle Publications.

30 Anon. (2003). *For the Safety of All: Images and Inspections of Irish Lighthouses*. Dublin. National Library of Ireland.

31 Taylor, R.M. (2004). *op. cit.*

32 Ward, N. & O'Brien, S. (2007). *Left for Dead: The untold story of the tragic 1979 Fastnet Race*. London. A. & C.B. Publishers.

33 Sharrock, J.T.R. (ed.) (1973). *The Natural History of Cape Clear Island*. Berkhamstead. T. & A.D. Poyser.

34 Wing, S. (2020). *The Natural History of Cape Clear 1959–2019*. Skibbereen. Privately published.

35 Somerville-Large, P. (1972). *op. cit.*

West Coast

1 Farrington (1954). *op. cit.*

2 Dillon, P. (1999). *Irish Coastal Walks*. Milnthorpe, Cumbria. Cicerone Press.

3 Clarke, A. (2014). *Walking the Sheep's Head Way.* Dublin. WildWays Press.

4 Sheehan, S. (2007). *Jack's World: Farming on the Sheep's Head Peninsula, 1920–2003.* Cork. Atrium Press.

5 Mee, A. *et al.* (2016). Reintroduction of White-tailed eagles *Haliaeetus albicilla* to Ireland. *Irish Birds* 10: 301–314.

6 Praeger, R.L. (1949). *op. cit.*

7 Cummins, S. (2019). *op. cit.*

8 O'Crohan, T. (1934) *The Islandman.* Dublin. Talbot Press.

9 Lavelle, D. (1976) *Skellig: Island Outpost of Europe.* Dublin. O'Brien Press.

10 Higgs, K.T. (2021). Tetrapod Trackway, Valentia Island, County Kerry. In: *The Coastal Atlas of Ireland.* (Devoy *et al.* eds.) Cork. Cork University Press.

11 Mitchell, F. (1990). *The Way that I Followed: A naturalist's journey around Ireland.* Dublin. Country House.

12 Mitchell, F. (1989). *Man & Environment in Valentia Island.* Dublin. Royal Irish Academy.

13 Smith, M. (2000). *An Unsung Hero: Tom Crean – Antarctic Survivor.* Cork. Collins Press.

14 Mac Conghail, M. (1987). *The Blaskets: a Kerry island library.* Dublin. Country House.

15 Mason, T.H. (1936). *The Islands of Ireland.* Cork. Mercier Press.

16 Praeger, R.L. (1912). Notes on the flora of the Blaskets. *Irish Naturalist* 21: pp. 157–163.

17 Turle, W.H. (1891). A visit to the Blasket Islands and the Skellig Rocks. *Ibis* Series 6, 3: pp. 1–12.

18 Brazier, H. & Merne, O.J. (1988). Breeding seabirds on the Blasket Islands, Co. Kerry. *Irish Birds* 4: pp. 43–64.

19 Trewby *et al.* (2006). *op. cit.*

20 Devoy, R., Cummins, V., Brunt, B., Bartlett, D. & Kandrot, S. (eds.) (2021). *The Coastal Atlas of Ireland.* Cork. Cork University Press.

21 O'Crohan (1934) *op. cit.*

22 Severin, T. (1978). *The Brendan Voyage.* London. Hutchinson.

23 Beebee, T.J.C. (2002). The Natterjack toad *(Bufo calamita)* in Ireland: current status and conservation requirements. *Irish Wildlife Manuals No. 10.* Dublin. Dúchas the Heritage Service.

24 King, J.L., Marnell, F., Kingston, N., Rosell, R., Boylan, P., Caffrey, J.M., FitzPatrick, Ú., Gargan, P.G., Kelly, F.L., O'Grady, M.F., Poole, R., Roche, W.K. & Cassidy, D. (2011). *Ireland Red List No. 5: Amphibians, Reptiles & Freshwater Fish.* Dublin. National Parks and Wildlife Service, Department of Arts, Heritage and the Gaeltacht.

25 Hill, J. (2009). *In Search of Islands: A life of Conor O'Brien.* Cork. Collins Press.

26 O'Sullivan, A. (2001). *Foragers, Farmers and Fishers in a Coastal Landscape.* Discovery Programme Monograph No.5. Dublin. The Royal Irish Academy.

27 Praeger, R.L. (1896). Kilkee. *The Irish Times*, 12 May 1896.

28 Foster, S., Boland, H., Colhoun, K., Etheridge, B. & Summers, R. (2010). Flock composition of purple sandpipers *Calidris maritima* in the west of Ireland. *Irish Birds* 9: 31–34.

29 Whilde, T. (no date). *Pocket Guide to the Cliffs of Moher*. Belfast. Appletree Press.

30 D'Arcy, G. (2017). *The Breathing Burren.* Cork. The Collins Press.

31 Cabot, D. & Goodwillie, R. (2018). *The Burren.* London. William Collins.

32 Heaney, S. (1996). 'Postscript', from *The Spirit Level.* London. Faber and Faber.

33 O'Rourke, C. (2006). *Nature Guide to the Aran Islands.* Dublin. Lilliput Press.

34 Roden, C. (1994). The Aran Flora. In: *The Book of Aran* (eds. J. Waddell, J.W. O'Connell & A. Korff). Kinvara. Tír Eolas.

35 Curtis, T.G.F., McGough, H.N. & Wymer, E.D. (1988). The discovery and ecology of rare and threatened arable weeds, previously considered extinct in Ireland, on the Aran Islands, County Galway. *Irish Naturalists' Journal* 22: pp. 505–512.

36 Robinson, T. (1986). *Stones of Aran: Pilgrimage.* Dublin. Lilliput Press in association with Wolfhound Press.

37 Scott, R.J. (2004). *The Galway Hookers: Sailing work boats of Galway Bay* (4th edition). Limerick. A.K. Ilen.

38 Nairn, R.G.W. (2005). The use of a high tide roost by waders during engineering work in Galway Bay, Ireland. *Irish Birds* 7: pp. 489–496.

39 Praeger, R.L. (1896). By the Western Ocean: a holiday in Connemara. *The Irish Times*, 12 May 1896.

40 Robinson, T. (1996). *Setting Foot on the Shores of Connemara and other writings*. Dublin. The Lilliput Press.

41 Mac an Iomaire, S. (2000). *The Shores of Connemara.* Translated and annotated by Padraic de Bhaldraithe. Kinvara. Tír Eolas.

42 Bassett, J.A. & Curtis, T.G.F. (1985). The nature and occurrence of sand-dune machair in Ireland. *Proceedings of the Royal Irish Academy* 85B, pp. 1–20.

43 Suddaby, D., Nelson, T. & Veldman, J. (2010). Resurvey of breeding wader populations of machair and associated wet grasslands in north-west Ireland. Irish Wildlife Manuals, No. 44. Dublin. National Parks and Wildlife Service, Department of the Environment, Heritage and Local Government.

44 Suddaby, D., O'Brien, I., Breen, D. & Kelly, S. (2020). A survey of breeding waders on machair and other coastal grasslands in Counties Mayo and Galway. Irish Wildlife Manuals, No. 119. Dublin. National Parks and Wildlife Service, Department of Culture, Heritage and the Gaeltacht.

45 Praeger, R.L. (1915). Clare Island Survey: General introduction and narrative. *Proceedings of the Royal*

Irish Academy 31 (1911–1915), section 1, part 1.

46 Feehan, J. (2019). *Clare Island.* Dublin. Royal Irish Academy.

47 Ferriter, D. (2018). *On the Edge: Ireland's offshore islands: a modern history*. London. Profile Books.

48 Hutchison, S. (2013). *op. cit.*

49 McNally, K. (1976). *The Sun-fish Hunt.* Blackstaff Press. Belfast.

50 Clarke, Jane (2021). At Purteen Harbour. In: *Divining Dante* (Edited by Paul Munden and Nessa O'Mahony). Recent Works Press. Woden ACT. Australia.

51 Cabot, D, Cabot, R. & Viney, M. (2020). The breeding seabirds and other birds on the Bills Rocks, County Mayo. *Irish Birds* 42: pp. 55–62.

52 McGreal, E. (2013). Probable breeding by Leach's Storm Petrel *Oceanodroma leucorhoa* at Bills Rocks, County Mayo with results of a census of its breeding birds. *Irish Birds* 9: pp. 636–638.

53 Scally, L., Pfeiffer, N. & Hewitt, E. (2020). The monitoring and assessment of six EU Habitats Directive Annex I Marine Habitats. *Irish Wildlife Manuals*, No. 118. Dublin. National Parks and Wildlife Service, Department of Culture, Heritage and the Gaeltacht.

54 Classen, R. (2020). *Marine Protected Areas – Restoring Ireland's Ocean Wildlife II. Report on Ireland's Failure to Protect Marine Natura 2000 Sites*. Dublin. Irish Wildlife Trust.

55 Ferriter, D. (2018). *op. cit.*

56 Nairn, R.G.W. & Sheppard, J.R. (1985). Breeding waders of sand dune machair in north-west Ireland. *Irish Birds* 3: pp. 53–70.

57 Dornan, B. (2000). *Mayo's Lost Islands: The Inishkeas.* Dublin. Four Courts Press.

58 Fairley, J. (1981). *Irish Whales and Whaling.* Belfast. Blackstaff Press.

59 Ottway, C. (1841). *Sketches in Erris and Tyrawly.* Dublin. W. Curry.

60 Ussher, R.J. & Warren, R. (1900). *The Birds of Ireland.* London. Gurney and Jackson.

61 Gange, D. (2019). *The Frayed Atlantic Edge: A historian's journey from Shetland to the Channel.* London. William Collins.

62 Hutchinson, C.D. (1989). *Birds in Ireland.* Calton. T. & A.D. Poyser.

63 Summers, C.F., Warner, P.J., Nairn, R.G.W., Curry, M.G. & Flynn, J. (1980). An assessment of the status of the common seal *Phoca vitulina vitulina* in Ireland. *Biological Conservation* 17: pp. 115–123.

64 Doyle, S., Walsh, A., McMahon, B.J. & Tierney, T.D. (2018). Barnacle Geese *Branta leucopsis* in Ireland: results of a census in 2018. *Irish Birds* 11: pp. 23–28.

65 Cowell, J. (1989). *Sligo: land of Yeats' desire.* Dublin. O'Brien Press.

66 O'Sullivan, A. & Breen, C. (2007). *op. cit.*

67 Heraughty, P. (1982). *Inishmurray: ancient monastic island.* Dublin. O'Brien Press.

68 Cabot, D.B. (1962). An ornithological expedition to Inishmurray, Co. Sligo. *Irish Naturalists' Journal* 14: pp. 59–61.

69 Cotton, D. (1982). Natural History and Conservation. In: *Inishmurray: ancient monastic island (Heraughty)*. Dublin. O'Brien Press.

70 Brown, B. (2011). *op. cit.*

71 Nairn, R. (2007). *A nature guide to Ardara-Portnoo area of County Donegal*. Wicklow. Nature Press.

72 Nairn, R.G.W. & Sheppard, J.R. (1985). *op. cit.*

73 Ferriter, D. (2018). *op. cit.*

74 Gibbon, M. (1935). *The Seals*. Dublin. Allen Figgis & Co.

75 Nelson, E.C. (2000). *Sea Beans and Nickar Nuts: A handbook of exotic seeds and fruits stranded on beaches in north-western Europe*. London. Botanical Society of the British Isles

North Coast

1 Suddaby *et al.* (2020). *op. cit.*

2 Praeger, R.L. (1934). *The Botanist in Ireland*. Dublin. Hodges Figgis & Co.

3 Anon. (1934). *Press opinions: The Botanist in Ireland*. Pamphlet published by Hodges Figgis & Co. Dublin.

4 McErlean, T. (2011). The maritime heritage of Lough Swilly. In: *Lough Swilly: a living landscape* (ed. A. Cooper). Dublin. Four Courts Press.

5 Classen, R. (2020). *op. cit.*

6 Kochmann, J. (2012). *Into the Wild: Documenting and Predicting the Spread of Pacific* Oysters *(Crassostrea gigas) in Ireland.* PhD thesis. University College Dublin.

7 Whatmough, J. (2011). *op. cit.*

8 Ferriter, D. (2018). *op. cit.*

9 McNally, K. (1978). *The Islands of Ireland.* London. Batsford.

10 Long, B. (1993). *op. cit.*

11 MacPolin, D. (1999). *The Drontheim: Forgotten sailing boat of the North Irish Coast.* Moville, Co. Donegal. Privately published.

12 Carter, R.W.G. & Wilson, P. (1991). Chronology and Geomorphology of the Irish Dunes. In: *A Guide to the Sand Dunes of Ireland* (ed. M.B. Quigley). Galway. European Union for Dune Conservation and Coastal Management. pp: 18–41.

13 O'Sullivan, A. & Breen, C. (2007). *op. cit.*

14 Watson, P. (2012). *The Giant's Causeway and the North Antrim Coast.* Dublin. O'Brien Press.

15 Watson, P. (2012). *op. cit.*

16 Whatmough, J. (2011). *op. cit.*

17 Nairn, G. & Whelan, M. (1993). Around Ireland with Drama. *Irish Cruising Club Annual 1993.* Dublin. Irish Cruising Club Publications.

18 Jones, K.B. (ed.) (2020). *Northern Ireland Seabird Report 2019.* Thetford. British Trust for Ornithology & Northern Ireland Environment Agency.

19 Brown, B. (2011). *op. cit.*

Crossing the Bar

1 Robinson, T. (1996). *op. cit.*

2 Farrington, A. (1954). Robert Lloyd Praeger 1865–1953. *Irish Geography* 3 (1): pp. 1–4.

3 Lysaght, S. (1998). *op. cit.*

4 Praeger, R.L. (1953). *Irish Landscape.* Dublin. Sign of the Three Candles.

5 Correspondence in the Praeger Collection held in the Library of the Royal Irish Academy, Dublin.

6 Farrington, A. (1954a). *op. cit.*

Turning the Tide

1 W.J.C. (1881). A Trawling Excursion in Dublin. *The Irish Monthly*, Vol. 9 (97), pp. 371–374. Published by Irish Jesuit Province.

2 Kelly, F. (2020). *Common Fisheries Policy 2020: A discarded opportunity.* Kilcoole. BirdWatch Ireland.

3 Clarke, M., Farrell, E.D., Roche, W., Murray, T.E., Foster, S. & Marnell, F. (2016). *Ireland Red List No. 11: Cartilaginous fish [sharks, skates, rays and chimaeras].* Dublin. National Parks and Wildlife Service, Department of Arts, Heritage, Regional, Rural and Gaeltacht Affairs.

4 Pacoureau, N. *et. al.* (2021). Half a century of global decline in oceanic sharks and rays. *Nature* 589: pp. 567–571.

5 Alheit, J. & Hagen, E. (1997). Long-term climate forcing of European Herring and Sardine populations. *Fisheries Oceanography* 6, pp. 130–139.

6 Bolster, W.J. (2012). *The Mortal Sea: fishing the Atlantic in the age of sail.* Harvard University Press.

7 Classen, R. (2020). *op. cit.*

8 Viney, M. (2006). Sands of time are running out for shellfish. *The Irish Times.* 11 March 2006.

9 Boland, H. & Crowe, O. (2012). *Irish Wetland Bird Survey: waterbirds status and distribution 2001/02 – 2008/09*. Kilcoole. BirdWatch Ireland.

10 Gittings, T., Breffni M., O'Donoghue, P., Clarke, S. & Tully, O. (2016). *Winter storms: Oystercatcher population dynamics and feeding ecology in Dundalk Bay, Co. Louth* 2011/12–2014/15. International Wader Study Group Annual Conference 2016.

11 Burke, B., Lewis. L.J., Fitzgerald, N., Frost, T., Austin, G. & Tierney, T.D. (2018). Estimates of waterbird numbers wintering in Ireland 2011/12 – 2015/16. *Irish Birds* 41: pp. 1–12.

12 Wilkins, N.P. (2004). *op. cit.*

13 Dennis, J.H., Jackson, E. & Burke, B. (2020). *Annual Aquaculture Report* 2020. Dún Laoghaire. Bord Iascaigh Mhara.

14 Gittings, T. & O'Donoghue, P. (2012). *op. cit.*

15 Zwerschke, N., Kochmann, J., Ashton, E.C., Crowe, T.P., Roberts, D. & O'Connor, N.E. (2017). Co-occurrence of native *Ostrea edulis* and non-native *Crassostrea gigas* revealed by monitoring of intertidal oyster populations. *Journal of the Marine Biological Association of the United Kingdom* 98: pp. 1–10.

16 Christianen, M. J. A. (2018). Return of the native facilitated by the invasive? Population composition, substrate preferences and epibenthic species richness of a recently discovered shellfish reef with native European flat oysters *(Ostrea edulis)* in the North Sea. *Marine Biology Research* 14, pp. 590–597.

17 Cummins *et al.* (2019). *op. cit.*

18 Chivers, L.S., Lundy, M.G., Colhoun, K., Newton, S.F., Houghton, J.D.R. & Reid, N. (2012). Foraging trip time-activity budgets and reproductive success in the black-legged kittiwake. *Marine Ecology Progress Series* 456: pp. 269–277.

19 Cummins, S., Lewis, L.J. & Egan, S. (2016). *Life on the Edge – Seabirds and Fisheries in Irish Waters*. Kilcoole. BirdWatch Ireland.

20 Ryan, C., Berrow, S. D., McHugh, B., O'Donnell, C., Trueman, C. N. & O'Connor, I. (2014). Prey preferences of sympatric fin *(Balaenoptera physalus)* and humpback *(Megaptera novaeangliae)* whales revealed by stable isotope mixing models. *Marine Mammal Science* 30: pp. 242–258.

21 Molloy, J. (2006). *The Herring Fisheries of Ireland, 1900–2005: Biology, Research, Development, and Assessment.* Galway. Marine Institute.

22 Nairn, R., Jeffrey, D. & Goodbody, R. (2017). *op. cit.*

23 O' Sullivan, K. (2020). Swimming ban at three beaches following overflow at Ringsend wastewater treatment plant. *The Irish Times.* 19 June 2020.

24 EPA (2020). *Urban Waste Water Treatment in 2019*. Wexford. Environmental Protection Agency.

25 Wall, B., Cahalane, A. & Derham, J. (2020). *Ireland's Environment: An integrated assessment*. Wexford. Environmental Protection Agency.

26 Lusher, A.L., Hernandez-Milian, G., Berrow, S., Rogan, E. & O'Connor, I. (2017). Incidence of marine debris in cetaceans stranded and bycaught in Ireland: Recent findings and a review of historical knowledge. *Environmental Pollution* 232, pp. 467–476.

27 Acampora, H., Lyashevska, O., Van Franeker, J.A. & O'Connor, I. (2016). The use of beached bird surveys for marine plastic litter monitoring in Ireland. *Marine Environmental Research* 120, pp. 122–129.

28 Collins, M. *et.al*. (2013). Long-term Climate Change: Projections, Commitments and Irreversibility. In: *Climate Change 2013: The Physical Science Basis. Contribution of Working Group I to the Fifth Assessment Report of the Intergovernmental Panel on Climate Change*. Cambridge. Cambridge University Press.

29 Pilkey, O.H., & Cooper, J.A.G. (2014). *The Last Beach*. Durham and London. Duke University Press.

30 Pilkey, O.H. & Cooper, J.A.G. (2014). *ibid*.

31 DELG / Department of the Environment and Local Government (2001). *Coastal Zone Management*. Spatial Planning Unit, Dublin.

32 Devoy, R. (2003). Coastal Erosion. In: B. Lawlor, ed. 2003. *The Encyclopaedia of Ireland*. Gill and McMillan Ltd, Dublin. pp. 215–216.

33 Pilkey, O.H., Pilkey-Jarvis, L. & Pilkey, K.C. (2016). *Retreat from a Rising Sea: Hard decisions in an age of climate change*. New York. Columbia University Press.

34 Pilkey *et al.* (2014). *op. cit.*

35 Fogarty, P. (2017). *op. cit.*

36 Marine Protected Area Advisory Group (2020). *Expanding Ireland's Marine Protected Area Network: A report by the Marine Protected Area Advisory Group*. Report for the Department of Housing, Local Government and Heritage, Ireland.

37 Classen, R. (2020). *op. cit.*

38 Nairn, R., Jeffrey, D. & Goodbody, R. (2017). *op. cit.*

39 Nairn, R. (2005). *op. cit.*

40 Dasgupta, P. (2021). *The Economics of Biodiversity: The Dasgupta Review*. London. HM Treasury.

Index